INTRODUCTION TO RF CIRCUIT DESIGN FOR COMMUNICATION SYSTEMS

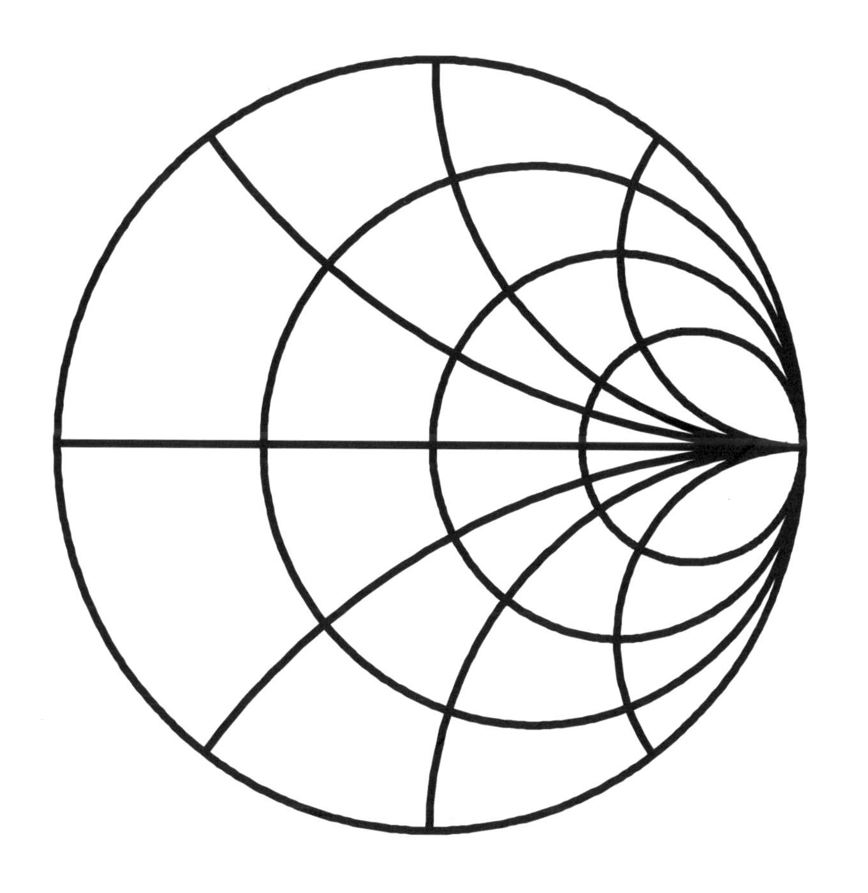

By

Roger C. Palmer

VE7AP

FOREWORD

This book originally started as the course notes for a class given in 1995 by the author for the Victoria-based West Coast Amateur Radio Association (WARA) entitled "Beyond The Advanced".

The course was designed to provide technical knowledge beyond that required for any amateur radio license, and provided information for those wishing to get extensively involved in "home-brewing" equipment, or those who just want to know more about how equipment works.

The original course notes have been expanded, and additional information has been added so that this book can be used as a “stand alone” reference book covering basic design techniques (from DC to UHF frequencies) for communication systems.

Much of the material touches on subjects normally taught at University level, but it is presented with a minimum of mathematics, such that anyone who has some experience with basic electronics should easily be able to absorb the material, even if they didn’t complete high school. No calculus is used, and only basic algebraic equations are presented.

Roger C. Palmer, P.Eng.
503 - 2605 Windsor Road
Victoria, BC V8S 5H9
Canada

ISBN 978-0-9950224-0-9

TABLE OF CONTENTS

INTRODUCTION

This book provides an insight into techniques that are commonly used in the design of modern RF communications equipment. Although the emphasis is on equipment or circuits that are part of communication systems, information is provided on a variety of general electronic design topics.

It is assumed that the reader has a general understanding of basic electronic concepts, such as that required to pass the U.S. General or the Canadian Advanced Amateur exam. No special mathematical skills should be necessary to make use of the material that is presented - basic Grade 10 algebra will be sufficient. No calculus will be used at any time. Some basic trigonometry is required in a few places, but a simple tutorial on the necessary concepts is provided in one of the Appendices.

This is not intended to be a formal text book with rigorous explanations, derivations, and difficult mathematics. It is assumed that the reader would prefer to get a good understanding of how circuits work, with just enough detail so that designs can be analyzed in a basic manner. Where appropriate, approximations and "rules of thumb" will be disclosed that can often simplify the design process. The book includes several design examples.

The reader should already be familiar with Ohms Law and reactance. Before proceeding further with this text, the reader might wish to refer to Appendix A, which is a summary of basic information that should have been encountered in one form or another in basic electronic training.

AC AMPLITUDE UNITS

Before starting to explore the fascinating world of RF design, we must first review the different types of units that are used to specify the amplitude of AC signals.

DC signals are easy to comprehend - the frequency is always zero, and there is only one type of voltage to specify. Usually a DC voltage will be stated as "DCV" or "Volts DC", but sometimes it will just be stated as "V" if it is obvious that the measurement is of a direct current voltage.

AC voltages are specified as either "peak-to-peak" (abbreviated as p-p) or RMS (which stands for Root Mean Square). Peak-to-peak voltage is easy to measure on an oscilloscope: it is simply the difference in volts from the positive peak to the negative peak, as shown below:

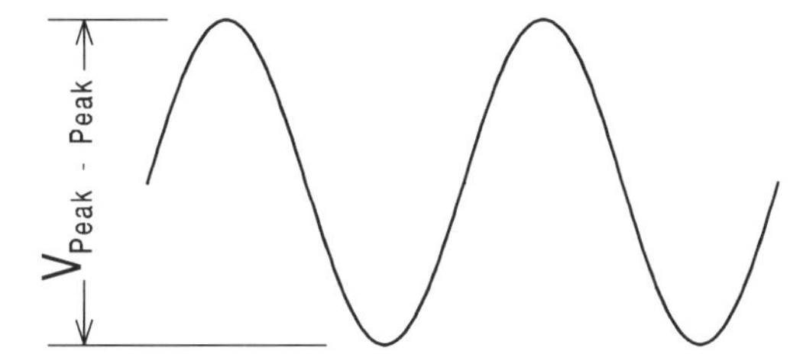

Although it is descriptive of an AC signal, peak-to-peak voltage measurements are ineffective when working with power measurements. A simple example will prove the point. If you think of the term "power" as referring to heating ability, it is easy to understand the concept. Imagine that a 10 VDC source is connected to a 1 ohm resistor that is immersed in a jar containing one pound of water. The power dissipated in the resistor is E^2/R, or 100 Watts. After the circuit has been connected for one minute, approximately 6 BTUs of heat will have been added to the water, and its temperature will have increased by 6° F. Now the experiment will be repeated, using a 10 volt peak-to-peak AC voltage across the same resistor. After one minute, the water's temperature has increased by less than one degree!

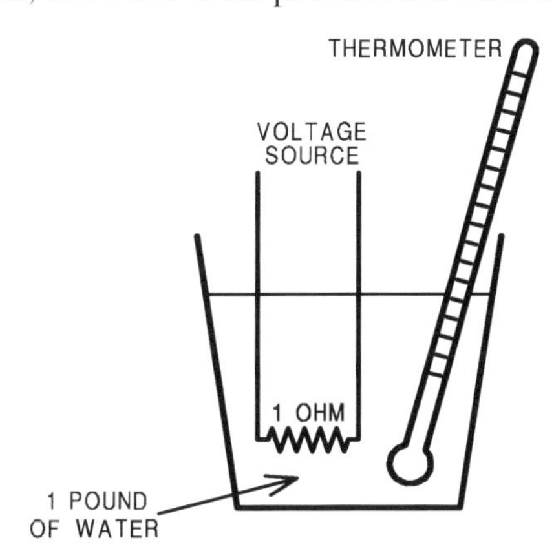

It is easy to imagine how the AC signal had less "heating capability" than the DC signal in these examples. Look at a complete cycle of the AC waveform and you will realize that the instantaneous voltage across the resistor has a maximum value of either plus or minus 10 volts (the resistor gets hot no matter what the polarity), but for much of the cycle, the instantaneous voltage is less than 10 volts. It therefore seems logical that the heating power will be less than when a constant 10 volts is applied to the resistor. In order to have the same "heating capability" as the 10 volt DC source, our AC signal will have to have a peak-to-peak voltage of 28.28.

The definition of an AC RMS voltage is such that it has the same "heating capability" as a corresponding DC voltage. Therefore, our 28.28 Vp-p signal can also be specified as 10 VRMS. For a sine wave, the relationship between peak-to-peak and RMS is $2\sqrt{2}$, or approximately 2.828. Mathematically, this relationship is derived by slicing up one cycle of the sine wave into a very large number of samples, and calculating the square of each of the instantaneous voltages. The resulting values are averaged, and then the square root is taken.

A square wave is a different issue. Here, the relationship between peak-to-peak and RMS voltage is a factor of 2 for a square wave with 50% duty cycle. For a rectangular pulse train that has a duty cycle of less than 50%, the factor will be smaller. The following diagram shows four different source voltages that can be connected to our water-filled "watt meter" and that will all have the same heating capability:

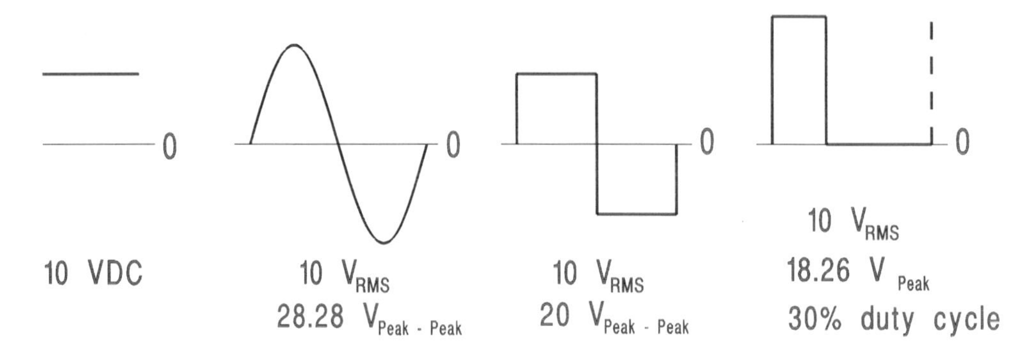

As mentioned before, it is easy to measure a signal's peak-to-peak voltage on an oscilloscope, although the accuracy will only be a few percent. If the signal being observed is a sine wave or symmetric square wave, it will then be possible to convert the reading to derive the RMS voltage if this is important to the application. If the waveform has an arbitrary shape, it is almost impossible to determine the RMS voltage by using an oscilloscope.

All DVM's (Digital Volt Meter) and VOM's (Volt Ohm Meter) have provision for measuring AC voltage. A few expensive models have what is called a "true RMS" capability - this means that they can accurately measure the RMS voltage of arbitrary waveforms. These instruments either have internal thermoelectric converters, or they use mathematical circuit elements to actually compute the RMS voltage.

Less expensive instruments simply rectify the AC signal with a basic diode circuit and then measure the resulting DC voltage, applying a scale conversion. In other words, the meter is actually measuring the peak voltage, <u>not</u> the actual RMS "heating capability". For sine waves, this method is perfectly acceptable, because the correct 2.828 conversion factor is built into the meter. For any waveform other than a sine wave however, the displayed value of AC voltage will be incorrect.

In RF design work, signal voltages are seldom given in terms of peak-to-peak voltage, RMS is almost always used. It is also common to express signal levels in terms of dBm (or dB relative to one milliwatt).

If dBm values are quoted, it is necessary to know the system impedance in order to convert to VRMS. Unless stated otherwise, assume that the system impedance is 50 ohms.

A signal having an amplitude of 0 dBm has a voltage such that it will dissipate 1 mW into a 50 ohm load. From Ohms Law, we can determine the corresponding RMS AC voltage as 0.224 V, or 224 mV.

A dB (deci-Bell or decibel) is a logarithmic measurement of a voltage or power ratio. The definition is:

For power ratios; $dB = 10 \log (P_2/P_1)$

For voltage ratios; $dB = 20 \log (V_2/V_1)$

If a circuit doubles the voltage (in other words, it amplifies the input signal by a factor of 2), Its voltage gain is 20 log (2) ≅ +6 dB. If the system impedance is the same for the input and output of the circuit, the power will have been increased by a factor of 4, and therefore the circuit's power gain is 10 log (4) ≅ +6 dB. Using the definition for voltage ratio in dB, it is now possible to convert any given AC voltage to dBm:

$dBm = 20 \log (V/0.224)$ (for a 50 ohm system)

The following table computes the equivalent power levels and voltages for several different values of dBm:

dBm	Power	Voltage
+20	100 mW	2.24 V
+10	10 mW	707 mV
+6	4 mW	448 mV
+3	2 mW	316 mV
0	1 mW	224 mV
-3	0.5 mW	159 mV
-6	0.25 mW	112 mV
-10	100 μW	70.7 mV
-20	10 μW	22.4 mV
-40	100 nW	2.24 mV
-60	1 nW	224 μV
-80	10 pW	22.4 μV

Measurements in dBm are usually based on a system impedance of 50 ohms.

In the cable TV industry it is common to use yet another unit of amplitude measure - the dB_{mV}. This is the relative level in dB of the voltage relative to 1 mV. This is quite easy to convert to actual voltage, because no power measurement is involved:

dB_{mV}	Voltage
+60	1 V
+40	100 mV
+30	31.6 mV
+20	10 mV
+10	3.16 mV
0	1mV
-10	316 μV
-20	100 μV
-30	31.6 μV
-40	10 μV
-60	1 μV

Note that cable TV measurements are usually made with a system impedance of 75 ohms.

THE TIME AND FREQUENCY DOMAINS

Most of us are used to viewing or imagining signals in the "time domain". When a signal is displayed on an oscilloscope it is presented in the time domain: the vertical axis represents voltage, and the horizontal axis represents time. A periodic waveform is one in which the voltage follows a specific repetitive pattern that repeats in time. As an example, an RF continuous wave carrier is a signal that follows a repetitive pattern of voltage as a function of time: the instantaneous voltage is proportional to the value of a trigonometric **sine** (abbreviated as "sin") function. If you are unfamiliar with the sine function, now might be a good time to review the Appendix section that provides an introduction to basic trigonometric concepts.

Ignoring phase for a moment, it is possible to describe a continuous AC signal of a single frequency as follows:

$V = A\sin(2\pi ft)$ where: f is the frequency in Hz.
t is the time in seconds
A is one half the peak-to-peak voltage

Note that the value of "sin" varies smoothly between +1 and -1 as the angle varies. You probably remembered angles as being measured in degrees (between 0 and 360), but it is common in electrical engineering to refer to angles using radian measure. A radian contains approximately 57 degrees, and there are exactly 2π radians in a complete circle. The term $2\pi f$ is referred to as the "angular frequency", and it written as the Greek letter "omega", which is ω. Angular frequency is measured in radians per second, as opposed to cycles per second (or "Hertz").

Imagine for a moment an AC signal with a 1 Hz frequency and a 2 volt peak-to-peak amplitude. At a given starting point, the time "t" will be zero, and therefore the value within the brackets will represent an angle of zero, and the instantaneous value of the signal will be zero. As the time increases, the angle increases. After 0.25 seconds, the angle will be $\pi/2$ radians (or 90 degrees), and the value of sin will be +1. The instantaneous voltage will therefore be +1 volts. As time goes on, the value of sin starts to decrease, until at 0.5 seconds, the angle is now π, and the voltage is again zero. As time goes on, the value of sin continues to decrease, until at an angle of $3\pi/4$ it has reached -1, and the instantaneous voltage is -1 volts. As time goes on, the value of sin again increases until it reaches zero at an angle of 2π radians after one second, and then the cycle continues all over again. This is illustrated in the following diagram:

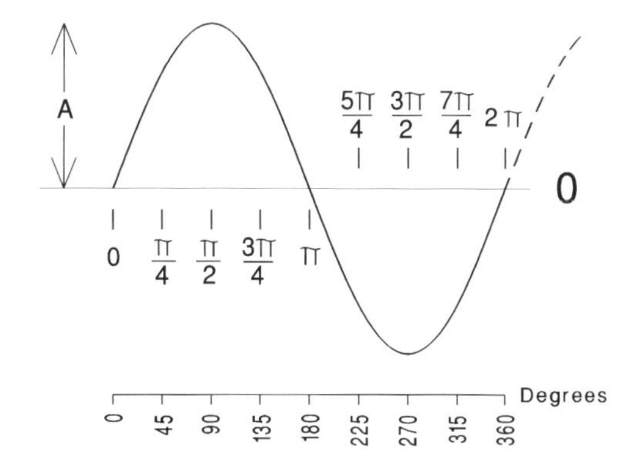

Note that in the above illustration, the angular measure is expressed both in radians (as a fraction of π) and in conventional degrees.

A signal whose instantaneous value can be described by a sine function is commonly referred to as a "sine wave".

We will now direct our attention to the "frequency domain". An instrument which displays signals in the frequency domain is called a **spectrum analyzer**. This instrument displays amplitude as a function of frequency. If our 1 Hz signal were connected to the input of a spectrum analyzer, we would see a single vertical line at the 1 Hz position. The height of the line is representative of the signal's amplitude. If a portion of the amateur 20 metre band were displayed on a spectrum analyzer, there would be many vertical lines next to each other, indicative of all the signals on different frequencies. Note that a spectrum analyzer provides amplitude information only - it does not give any data on phase.

A pure continuous sine wave is the only signal that has energy at a single frequency, and therefore will be represented by a single vertical line on a spectrum analyzer. If the signal departs at all from a perfect sine shape (in other words, it has some type of distortion), more than one frequency will be present: it will contain "harmonics", which are exact multiples of the fundamental frequency. If a cycle of the signal is non-symmetrical with respect to the vertical axis it will contain even harmonics; if the signal is symmetric it will contain odd harmonics.

Any imaginable continuous signal can be created by adding together harmonically-related sine waves of different frequencies and phase relationships. As an example, a perfect square wave of angular frequency ω can be expressed as follows:

$$V = A\left(\sin(\omega t) + \frac{1}{3}\sin(3\omega t) + \frac{1}{5}\sin(5\omega\, t) + \ldots + \frac{1}{N}\sin(N\omega t)\right)$$

where N is odd

This is equivalent to saying that a square wave is composed of a sine wave at its fundamental frequency, plus all of the odd harmonics, where the relative amplitude of each harmonic is equal to one over the harmonic number. It is because of this harmonic relationship that digital signals (which have quite sharp edges) tend to radiate so much RF noise. Note that there are no even harmonics in the power spectrum of a symmetrical (50% duty cycle) square wave. Indeed, there are no even harmonics from any waveform that has so-called "half wave symmetry", which means that for the function, $f(x) = -f(x+\pi)$. The following illustration approximates what would be seen on the screen of a spectrum analyzer when it is connected to a 1 MHz sine wave or a 1 MHz square wave:

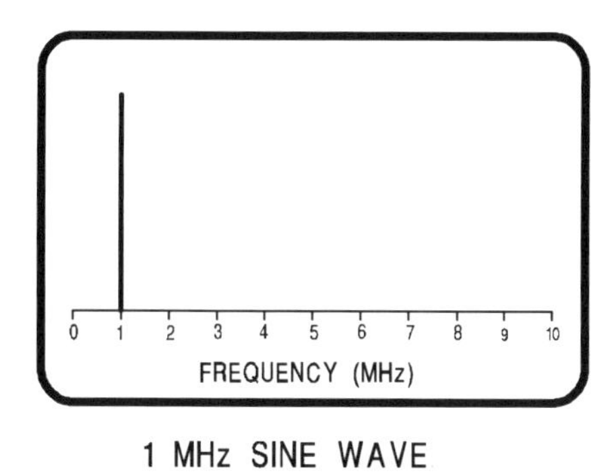

1 MHz SINE WAVE

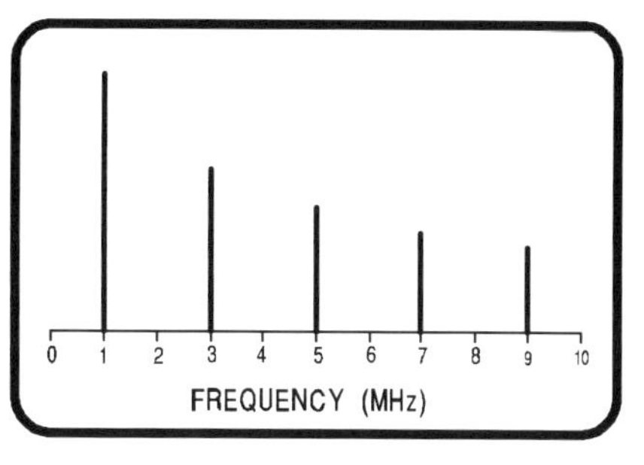

1 MHz SQUARE WAVE

In order to faithfully reproduce a 1 MHz perfect square wave, an oscilloscope would theoretically have to have an infinite bandwidth in order to capture the information from all of the harmonics.. In practice, as long as the scope's bandwidth is at least 10 times the fundamental frequency the displayed signal will look quite "square" to a human observer.

If a square wave is passed through a low pass filter, the higher harmonics will be attenuated. As a result, the signal will no longer look square, but will have more of a rounded shape

Any time that a voltage rapidly changes value (such as a logic level changing from one to zero or vice versa), energy is generated at quite high frequencies. The sharpness of a digital voltage transition is expressed by the "rise time" (or fall time), which is the time to go from 10% of the voltage swing to 90% of the final value. There is a relationship that expresses how much bandwidth is required to produce a given rise time (assuming a perfectly square starting waveform):

$$BW_{3dB} \approx \frac{0.35}{t_r}$$

where t_r. is the rise time (10% to 90%)
BW_{3dB} is the required bandwidth

A mathematician known as Fourier developed techniques for mathematically converting between the time domain and frequency domain. Using his non-trivial techniques, you can start with the mathematical description of a waveform in the time domain and derive an expression indicating the energy component at each of the component frequencies, as well as the phase relationship. This can be an involved and very tedious process. A technique that can be implemented successfully on computers is called the Fast Fourier Transform (or FFT for short). Test equipment designed to carefully analyze lower frequencies (such as audio) often uses an on-board computer to perform an FFT on the input signal in order to display the frequency domain information. These techniques are often used for vibration or seismic analysis.

Spectrum analyzers for RF are basically superheterodyne receivers which are repetitively tuned over the frequency of interest. At any given moment, the amplitude of the received signal is displayed on the vertical axis of the display. Rather than have a linear display of voltage on the vertical axis, it is more common to display amplitude in terms of dB_m. With a displayed dynamic range of over 60 dB, this allows distortion to be measured that would be totally undetectable on the time domain display of an oscilloscope.

There are other transformations apart from the Fourier transform. Hartley developed a transform based on square waves, and the new Wavelet approaches use yet another technique.

VOLTAGE AND CURRENT SOURCES

A perfect voltage source has zero internal impedance, and will deliver whatever current is necessary in order to deliver and maintain its selected voltage to the load.

A perfect current source has infinite internal impedance, and it will develop whatever voltage is necessary in order to supply its selected current to the load.

Sources are evaluated by examining their **load curves**. This is a graph which shows the interrelationship between the output voltage and the output current as the load is varied. The test circuit and load curves for these two types of sources are:

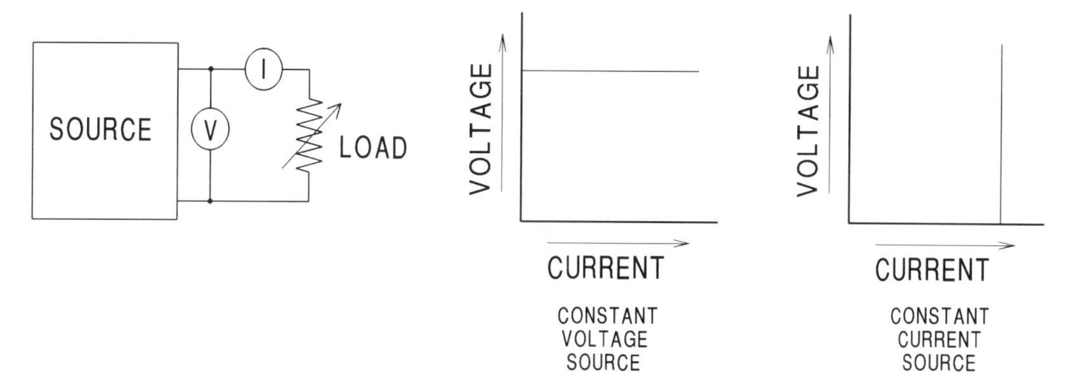

Real voltage sources (such as power supplies) have an internal resistance. This causes the output voltage to "droop" slightly as the load is increased. This is equivalent to having a small resistor (referred to as the "internal resistance") in series with a perfect voltage source. At some current, the internal resistance may appear to increase , as shown by the curvature on the right-hand end of the following load curve:

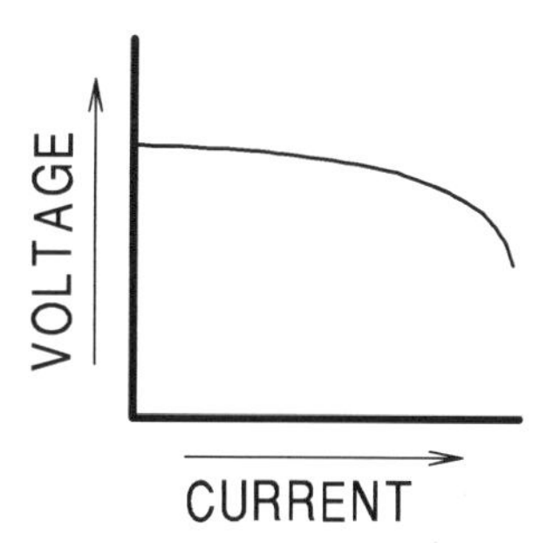

Note that the use of feedback can drastically reduce the perceived output impedance of a power supply.

A voltage source or current source can also be used as a "sink" if it can absorb current. As an example, a zener diode acts as voltage sink or "clamp". In the following discussions, we will often just refer to voltage or current sources, even if they are actually "sinks".

Thevenin Equivalent

If a resistor is in series with a perfect voltage source (or at least one having a very low internal resistance), the output will act just like a voltage source with an output resistance equal to the value of the resistor.

When a perfect voltage source (or at least one having a very low internal resistance) is connected to a resistive divider, the situation is only slightly more complicated. If no additional load is connected to the output of the resistive divider, the output voltage is equal to the voltage of the source times the resistive divide ratio. When current is drawn from the output of the resistive divider, the voltage falls as though the source has an internal resistance that is equal to the parallel resistance of the two resistors in the divider. The following figure illustrates this point:

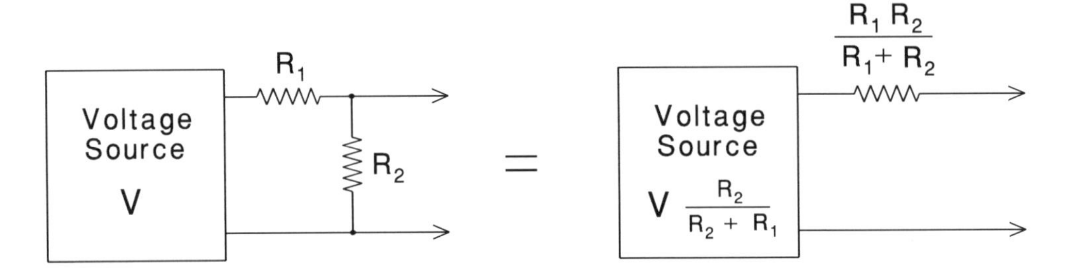

Note that this approach can be used to simplify schematics when you are trying to analyze a complicated circuit.

RC CIRCUITS

At low frequencies (less than 1MHz), the passive components in circuits are primarily either resistors or capacitors. Low frequency discrete amplifiers usually use capacitor coupling between stages unless the response is designed to go all the way down to DC. Let's examine a typical coupling network as might be used between two stages of transistor amplification. We will show both the circuit as it appears on the schematic, and the equivalent circuit that we will use to analyze the AC frequency response:

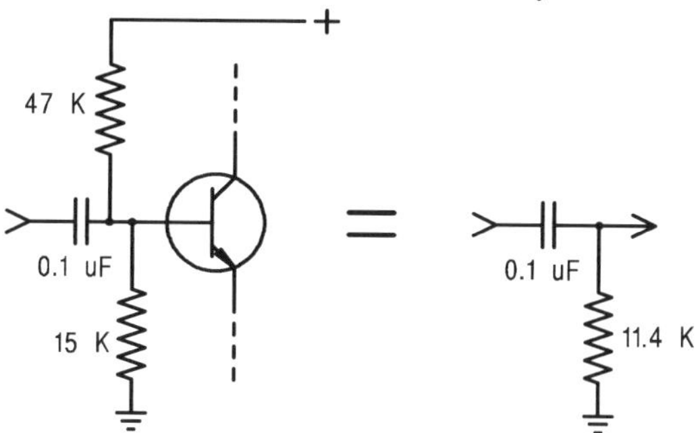

Notice that from an AC standpoint, we don't worry about the relative DC voltages as we examine this circuit: the positive power supply connection is the same as "ground" as far as AC signals are concerned. Assuming for the moment that the input impedance to the base of the transistor is very high, we can simplify the circuit by combining the two bias resistors into an equivalent single resistor whose value is the parallel combination of 47 K and 15 K.

The reactance of the capacitor at a given frequency can be determined from the formula $X_C = 1/2\pi fC$. An easier approach is to read the value directly off a reactance chart, such as the one in Appendix B. For higher AC frequencies, the capacitor's reactance is much smaller than the 11.4 K of effective load resistance, and AC signals are passed with virtually no attenuation. At lower frequencies the capacitor's reactance becomes significant, and response starts to fall off. At a frequency of 140 Hz, the reactance of the capacitor is equal to 11.4K, and you would normally expect the output voltage to be equal to one half of the input voltage (in other words, a throughput of 0.5, or 6dB of attenuation), but this simple assumption ignores the phase angle relationship of the circuits current. Later on in these book we will talk about "complex impedances", and the reader will learn that the attenuation at this point is actually 3dB, or a voltage ratio of 0.707.

As the frequency is decreased the reactance of the capacitor will increase further, and the attenuation will increase.

It is normal to specify the sine wave frequency responsc of an amplifier by the lowest and highest frequency that can be passed with less than 3dB of "roll off" from the peak response. For this coupling network, the 3dB low frequency limit is 140Hz. If it is desired to pass lower frequencies (such as all the way down to 20Hz for a hi-fi amplifier), larger capacitors can be used. As an example, increasing the value of the capacitor by a factor of 10 will move the low end down to 14Hz.

A 5KHz signal can be passed through this network without a problem, but if a 250 Hz square wave is passed through, the output voltage will exhibit significant "droop". The vertical parts of the square wave will be passed through the capacitor just fine, but the horizontal segments will decay back to the average value of the waveform. This is due to the time constant of the coupling network.

Time Constant

The following circuit illustrates the concept of RC time constant:

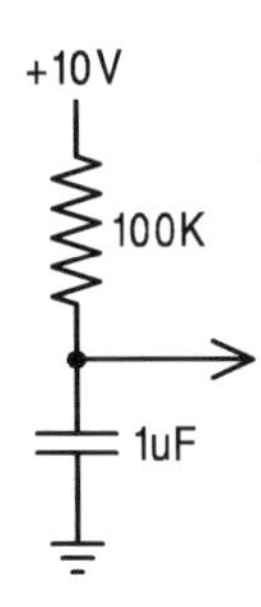

The capacitor is assumed to start with no charge. In other words, we will assume that there is zero voltage across the capacitor at time t = 0. As time progresses however, the capacitor will charge, and the output voltage will rise. The rate at which the voltage rises is a function of the circuit's time constant, which is defined as R times C and is measured in seconds. In this example, the time constant (product of R and C) is 0.1 seconds. The expression which describes the output voltage as a function of time is:

$$V_{out} = V_A\left(1 - e^{\frac{-t}{RC}}\right) = 10\left(1 - e^{\frac{-t}{0.1}}\right)$$

An actual plot of the output voltage as a function of time is shown below:

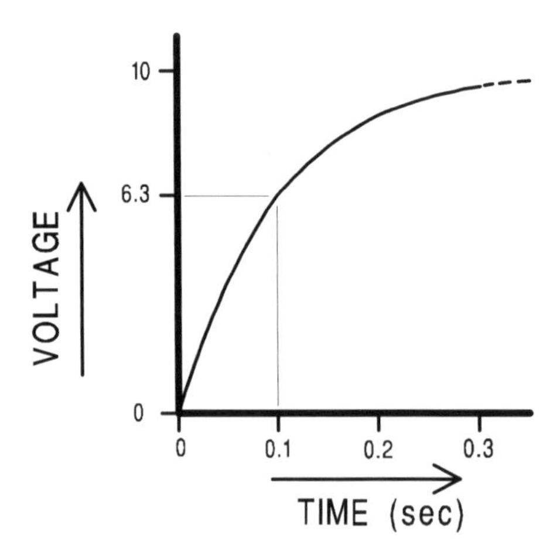

The **Time Constant** is simply the product of R times C. As you can see, the output voltage takes a very long time to reach the charging voltage (actually it never quite gets there). However, an easy rule of thumb to remember is that the voltage will rise to 63% of its final value after RC seconds.

A charged capacitor that is discharged through a fixed resistor also displays an exponential voltage curve. Assuming that the capacitor starts with an initial voltage of V_A volts, the output voltage as a function of time is described by:

$$V_{out} = V_A \, e^{\frac{-t}{RC}}$$

The rule of thumb here is that the output voltage will discharge to 37% of its initial value after RC seconds.

Taking another look at RC coupling, consider the following circuit:

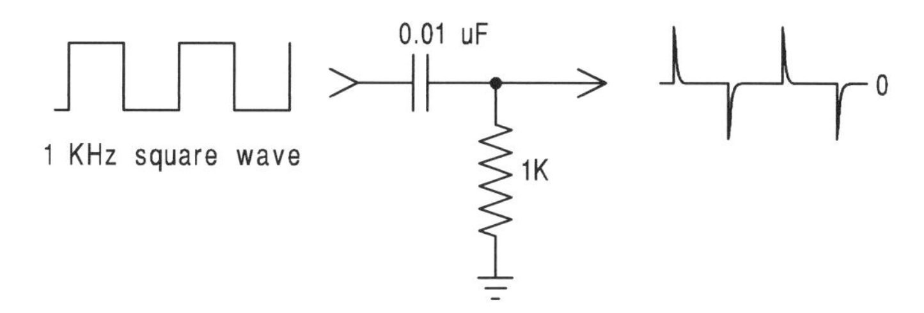

In this circuit, the RC time constant of the coupling components is 10 microseconds. The period of one half of a cycle of our 1KHz square wave is 500 microseconds, which is 50 times the time constant. As a result, the capacitor has lots of time to discharge almost completely after every vertical transition of the input waveform, and all that is left at the output is a series of "spikes". This is called a **differentiator** circuit, because the output is roughly proportional to the **rate of change** of the input waveform. If you look closely at this circuit, you can see that in the frequency domain, this is a **high pass filter**. Remembering that a square wave consists of a fundamental and all the odd harmonics, it can be seen that we are attenuating the fundamental and the low order harmonics, while the higher harmonics are able to pass through the circuit.

Now we will look at a simple RC **low pass filter**:

10K

100 Hz

Square wave input

10 KHz

0.1 uF

The RC time constant is equal to 1 millisecond. If a 100 Hz square wave is applied to the input, the output will look quite like a square wave, although the rise and fall times may be somewhat greater than those at the input, and a rounding of the corners will be apparent. However, if a 10KHz square wave is applied, the output will look like a slightly rounded triangular waveform. This can be understood by thinking in the time domain of simple charging and discharging action in the capacitor, or can be analyzed by realizing that this circuit is a low pass filter, and the higher order harmonics are being attenuated.

Through both of these examples, the student should be trying to understand the relationship between what is occurring in the time domain to what is happening in the frequency domain. It should also be noted that the use of square waves as a test signal can be very beneficial: by looking with an oscilloscope at the output of the circuit under test, it is easy to see if the circuit's frequency response extends appreciably higher and/or lower than the frequency of the test square wave signal.

LINEAR AMPLIFIERS (Small Signal)

The function of a small signal amplifier is to increase the amplitude of the input signal while preserving its waveshape. Note that in this chapter we will primarily be looking at signal voltage levels, not power levels. This chapter will also focus on low frequencies (less than 100 KHz), where device internal capacitances can be mostly ignored.

Tubes

The oldest form of amplifier makes use of vacuum tubes, usually abbreviated to just "tubes" (or "valves" if you live in the U.K.). The basic principal is similar to that exploited in other, more modern amplifying devices - a small signal is used to control the flow of current through a device. It is possible to think of a tube (or a FET) as a "valve" that controls the flow of current through the device.

Tubes are categorized based on the number of elements (apart from the heater). A triode is composed of the cathode, grid, and plate, while a pentode has five elements - the cathode, control grid, screen grid, suppressor, and plate. The plate is also sometimes referred to as the "anode". At low frequencies, tubes have very high input impedances. For small signal applications, triodes are commonly used at low frequencies, and pentodes at higher frequencies. A pentode's screen grid minimizes capacitance from the plate to the grid.

Tubes are characterized by a set of plate curves. This is a series of plots that indicates how much current flows from the cathode to the plate (electron flow) as a function of the plate-to-cathode voltage for a given grid-to-cathode voltage. A typical set of plate curves for a triode and a pentode are shown below. Note that the grid is biased negatively with respect to the cathode for most of the useable operating zone.

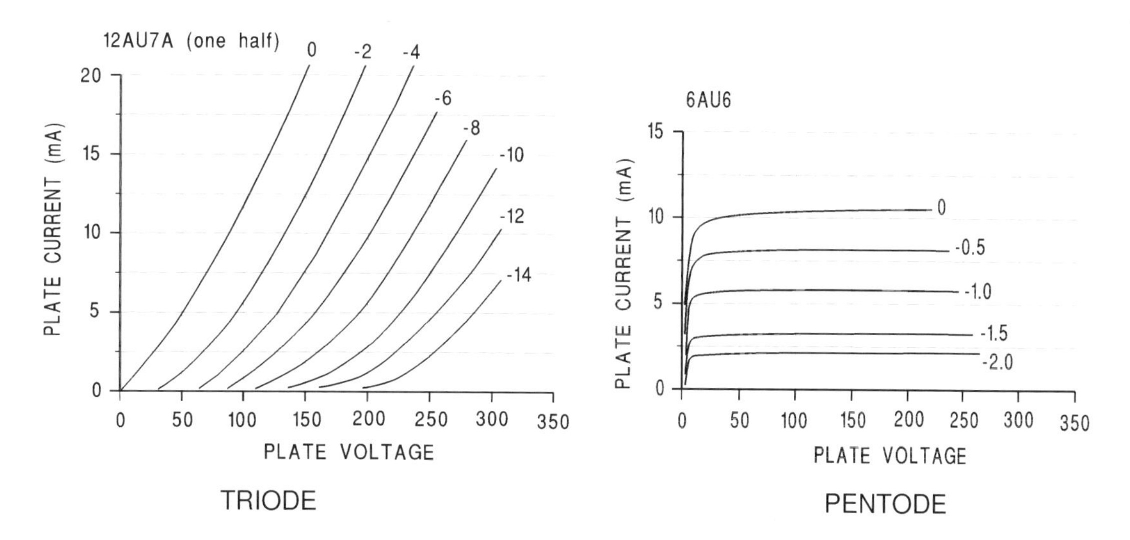

Note that the above plate curves plot plate voltage along the horizontal axis and plate current on the vertical axis - this is opposite from the convention used for "load curves". It should be apparent that the plate curves for the pentode approximate those for a constant current source for plate voltages of more than about 20 VDC.

The negative bias on the grid can be generated either by a separate supply (not very convenient), or by using a cathode bias resistor, as shown in the following example using one half of a 12AU7 triode:

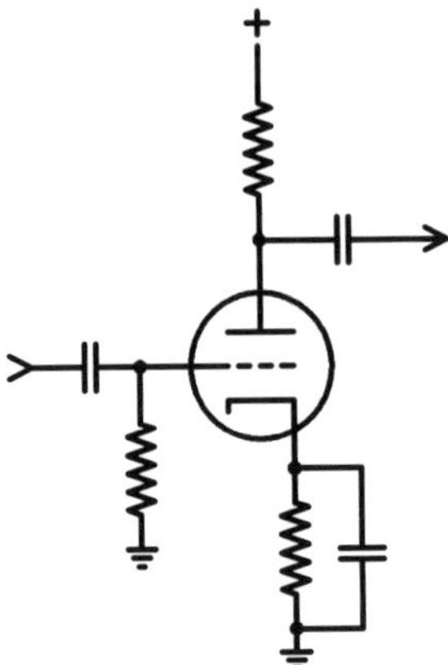

In order to prevent "degeneration" from the cathode resistor, it should be bypassed with a capacitor which has a low reactance at the lowest frequency to be amplified. The grid leak resistor can be quite large (1MΩ is a typical value), and this allows small coupling capacitor values to be used.

The following is a "quick and dirty" way to select the values for a low frequency AC-coupled triode amplifier such as that shown above:

1. Select the power supply voltage. Usually this is given, but if it is not, choose a voltage that is well below the tube's maximum.
2. Select a quiescent operating current. In the absence of any other compelling reason, a value of 1 or 2 mA is often ideal.
3. Choose the plate resistor such that its voltage drop when passing the selected operating current is slightly less than (or equal to) one half of the supply voltage.
4. Draw a load line on the plate curve for the tube. This is done by drawing a line between a point on the horizontal axis at the supply voltage, to a point on the vertical axis that is equal to the supply voltage divided by the plate resistor.
5. Find the point on the load line that represents the selected operating current, and determine what grid voltage this relates to.
6. Divide the required grid voltage by the selected operating current to determine the value of the cathode resistor.
7. Choose a value of cathode bypass capacitor such that its reactance at the lowest frequency of interest is no more than 100 ohms.
8. Set the grid resistor to 1 M.

Considering a small signal, low frequency, class A tube amplifier, we can now talk about some of the factors that affect the performance.

A tube's amplification factor μ is the ratio of the plate voltage change required to maintain a constant plate current to a grid voltage change which caused the change in plate current. The amplification factor for triodes varies from about 3 to 300.

A tube's plate resistance r_p is the reciprocal of the slope of the plate curve at a given grid voltage. Note that the plate resistance of pentodes is much higher than that for triodes. This means that for a pentode with a constant grid voltage, the plate current does not vary much with a change in plate voltage: in other words, it acts like a current source.

If the tube's plate voltage is held constant, the amount of change in plate current for a given change in grid voltage is called the transconductance g_m. The three tube parameters are inter-related as follows:

$$g_m = \frac{\mu}{r_p}$$

Tubes are primarily used these days for high power and/or specialized high frequency operation. A large proportion of satellites in orbit today use tubes! (Travelling Wave Tubes)

FETs

FETs have many electrical similarities with tubes. They have very high input impedances, and the device acts quite like a controllable current source. The current which flows from the source to the drain (assuming electron flow in an N-type device) is controlled by the voltage between the gate and the source. The following set of curves shows the drain current I_D as a function of the drain-to-source voltage V_{DS} for a number of different gate-to-source voltages V_{GS} for a typical N-type FET:

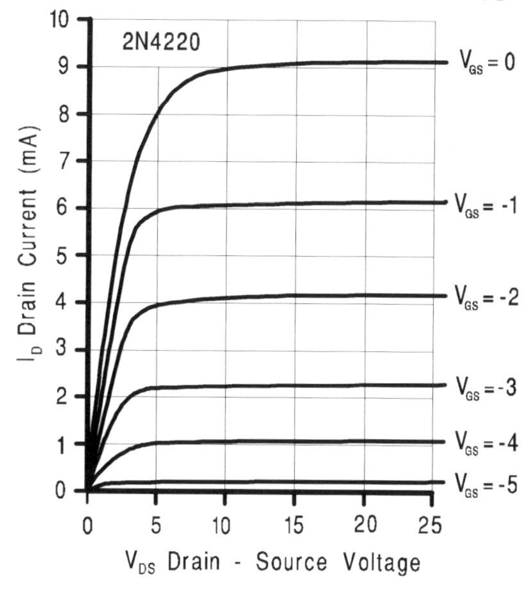

Note the "flatness" of the curves: these devices have a very high output impedance, and act as good current sources once V_{DS} exceeds a volt or so. This particular device is a so-called "depletion" type: it is normally biased with the gate negative with respect to the source: it behaves quite similarly to a pentode. Another common type is the "enhancement-mode" FET, which requires a positive V_{GS} in order to get any current to flow at all.

The voltage between the gate and the source at the point that drain current stops flowing is called the "pinch off voltage", and it is written as "V_P". Another term that is used to characterize the DC performance of an FET is "I_{DSS}", which is the current that flows from source to drain when the gate to source voltage is zero. Note that for small values of V_{DS}, the FET acts as a controllable resistor: this characteristic is often used in voltage variable attenuators, variable gain stages, or AGC circuits.

FETs can be either N-channel or P-channel. There are junction FETs (J-FETs) and MOSFETs (Metal Oxide Semiconductor FETs). Some MOSFETs have two gates, allowing two different signals to effect the drain current: this is somewhat analogous to multi-grid tubes, and this type of device is often used in mixers.

The actual drain current of an FET is a function of V_{GS}, V_P, and I_{DSS}:

$$I_D = I_{DSS}\left(1 - \frac{V_{GS}}{V_P}\right)^2$$

where I_{DSS} = drain current when $V_{GS} = 0$
V_{GS} = gate to source voltage
V_P = pinch off voltage

Depletion-mode FETs used in linear class A amplifiers are biased in a similar fashion to tubes. It is common to place a high impedance from the gate to ground (just like a tube's grid leak resistor), and to use a source resistor to bias the source at a positive voltage with respect to the gate. The source resistor should be bypassed at signal frequencies with a capacitor, otherwise degenerative (negative) feedback will result. A typical low frequency amplifier and its equivalent circuit are shown below:

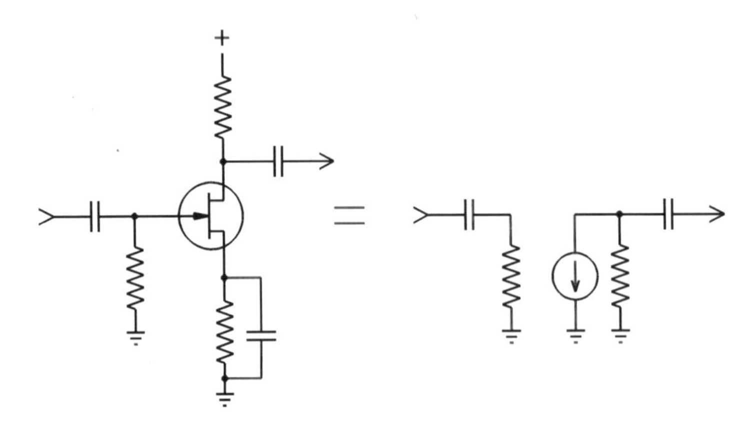

Note that the equivalent circuit shown above is valid only for frequencies low enough that the FET's internal capacitances can be ignored. It also assumes that the frequency is high enough such that the reactance of the source resistor's bypass capacitor is very small. The value of the current source is a function of the voltage between the gate and the source.

FETs make very good switches for both analog and digital signals. When turned "on", they exhibit a low resistance (often much less than an ohm). For switching analog signals, a popular device is a "transmission gate" such as a 4016 or 4066: this IC contains 4 analog switches, each of which is a P-channel and an N-channel device in parallel.

FETs are often used as the input stage in amplifier circuits because of their high input impedance. FETs are available for use in small signal or power amplifier applications for frequencies from audio through to the microwave range.

Bipolar Transistors

Bipolar transistors are either NPN or PNP. The basic material that the device is manufactured from is silicon for most general purpose applications, although Gallium Arsenide is used for some special UHF/microwave amplifiers.

When used in its most normal form (Class A, common emitter), the base to emitter junction is forward biased, therefore it acts similarly to a conducting diode. The base voltage is therefore about 0.7 volts higher than the emitter voltage for an NPN transistor. Being a forward-biased junction, the base-to-emitter voltage V_{BE} decreases approximately 2 mV per degree C.

The collector is reverse-biased from the other two terminals. A small leakage current called I_{CBO} flows from the collector to the base. And its value approximately doubles for every 10 degrees C of temperature rise in the junction area. In high power designs, this leakage current can lead to thermal runaway unless a low impedance path is provided from the base.

For low frequencies, a transistor can be considered as a controllable current source. The current source is controlled by the current flowing from the base to the emitter. At low frequencies, the ratio of current flow in the base to current flow in the collector is commonly called beta (β) or h_{FE}. Beta is commonly in the range of 50 to 200, and it is quite temperature dependent. Beta increases with increasing temperature. Typical curves (NPN transistor) of collector current as a function of V_{CE} are shown below for a variety of base currents:

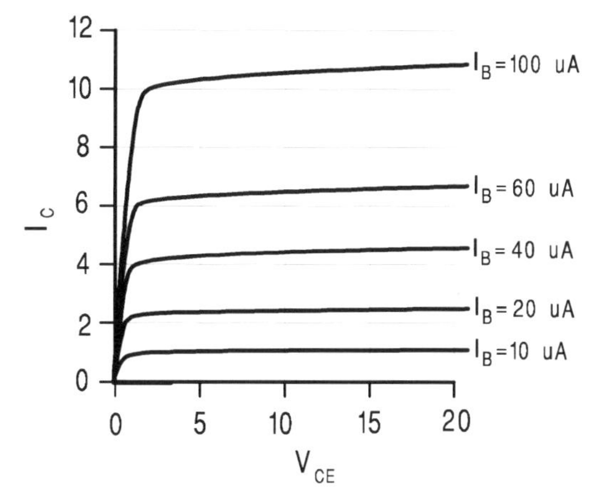

Note that in the above curves, the base-to-emitter voltage V_{BE} is about 0.65 to 0.70 volts all the time that the transistor is operating in its linear region. The important parameter which controls the collector current is the current through the base, **not** the voltage. From a DC standpoint, the collector current $I_C=\beta I_B$, or $I_C=h_{FE}I_B$. Because the voltage from the base to the emitter is fairly constant, it should be intuitive that the input impedance of a transistor amplifier is much less than a FET or tube. Indeed, from the standpoint of very low frequencies, the connection from the base to the emitter looks just like a forward biased diode. Two low frequency equivalent circuits for a bipolar transistor are shown below:

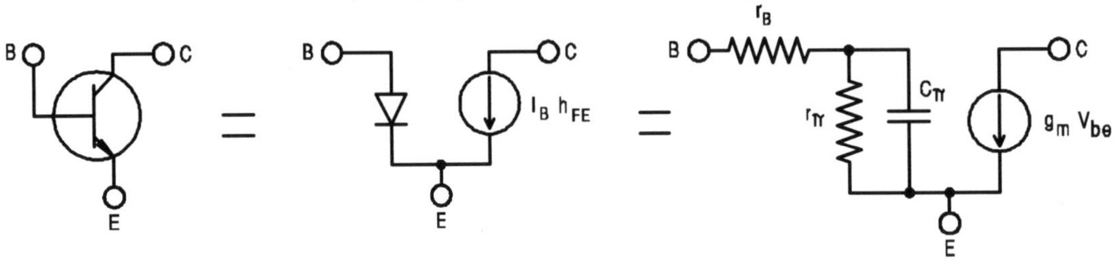

The middle equivalent circuit is useful for understanding how the device is biased, whereas the right hand circuit (the so-called "Hybrid π" model) is more suited for actually analyzing small signal AC circuit performance because it includes some frequency-dependent effects. The Hybrid π model is a useful way of analyzing small signal transistors in circuits operating at frequencies of less than about 10MHz. Note that the Hybrid π model assumes that the input signal is a voltage, not a current. The resistor r_B represents the contact resistance to the base region, and is often about small enough that it can be ignored at low frequencies. There is also a small capacitance between the collector and the base, but we will also ignore it at low frequencies. The parameters of this model are calculated as follows:

$$r_\pi \cong (\beta+1)\frac{25}{I_C},$$ where I_C is measured in milliamps.

$$C_\pi \cong \frac{\beta}{2\pi f_T r_\pi},$$ where f_T is the transistor's "transition frequency" (from the data sheet).

$$g_m \cong \frac{I_C}{25},$$ where I_C is measured in milliamps.

Using a very gross approximation, the gain of an amplifier will typically fall to less than unity for an input frequency of f_T. Note that the model as we have drawn it does not include the effects of lead inductance or capacitance between the collector and the base and emitter. Different kinds of models are used to describe the behaviour of transistors at high frequencies or in high power situations.

The input resistance seen at the base of a transistor amplifier can be approximated by multiplying β times the sum of the unbypassed emitter resistor and the operating current (in mA) divided into 25. As an example, consider the case of a transistor with a β of 150, an emitter resistor of 82 ohms, and an operating current of 2 mA. Assuming that the frequency is such that any emitter bypass capacitor is effectively out of the circuit, the input resistance observed looking into the transistor's base will be:

$$150\left(82+\frac{25}{2}\right)=14\,K\Omega$$

The fact that the DC base voltage is always about 0.7 volts above the emitter voltage can be used beneficially in several ways.

If it is required to produce a buffer amplifier with no voltage gain, but very low output impedance, an **"emitter follower"** can be used. The signal is applied to the base, and the output is taken from the emitter: the output impedance will roughly be equal to the input impedance divided by h_{FE}, and the output voltage will quite faithfully follow the input voltage, except that it will be one V_{BE} lower.

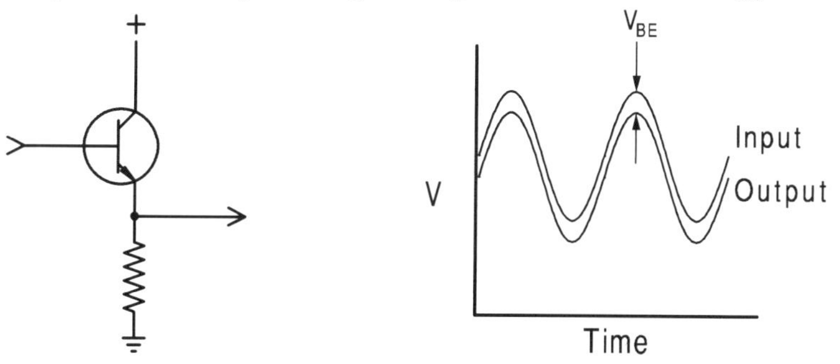

If a fixed DC voltage is applied to the base, and a fixed resistor is connected between the emitter and ground, then the collector terminal will act as though it is connected to a constant current sink. The value of the current sink is equal to the base voltage minus one V_{BE}, divided by the emitter resistor's value.

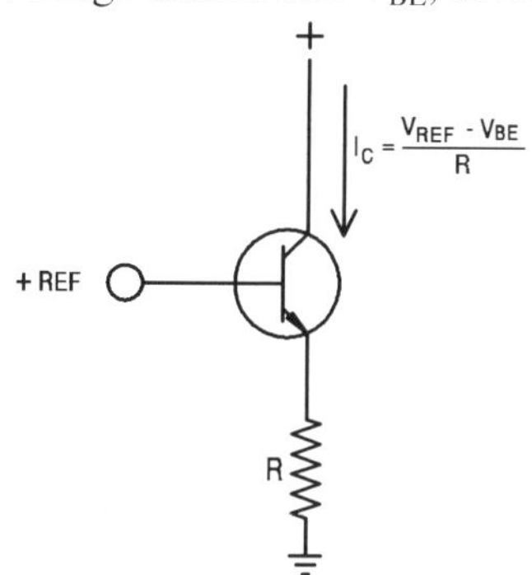

Biasing a bipolar transistor for linear amplification consists of configuring the device so that the quiescent collector current is maintained at the desired value despite variations in temperature. Recall that h_{FE} increases with temperature, and so does the leakage current from the collector to the base. The value of V_{BE} decreases with temperature. For this reason, it is impractical to try to bias a transistor by connecting its emitter to ground, and adjusting a voltage source on its base. Use some amount of emitter resistance to form an approximation of a constant current source, and use a resistive voltage divider to place a constant voltage on the base. The parallel resistance of the two base resistors must not be so low that they "swamp" the input signal, or so high that they are unable to "bleed away" leakage current from collector to emitter. Don't set the bias so that a large voltage is dropped across the emitter resistor: this will just detract from the available output voltage swing. Don't use too small a voltage across the emitter resistance, or the bias current will change significantly as V_{BE} varies with temperature. Remember that the important temperature is the junction temperature (not the outside air temperature), and that this will reflect the amount of power dissipated in the transistor.

Let's do a simple example. Assume that it is desired to build a low level audio amplifier using a single NPN transistor that operates from a 10 volt supply. The chosen transistor has an h_{FE} of 100, and an f_T of 200 MHz. We will use the following circuit:

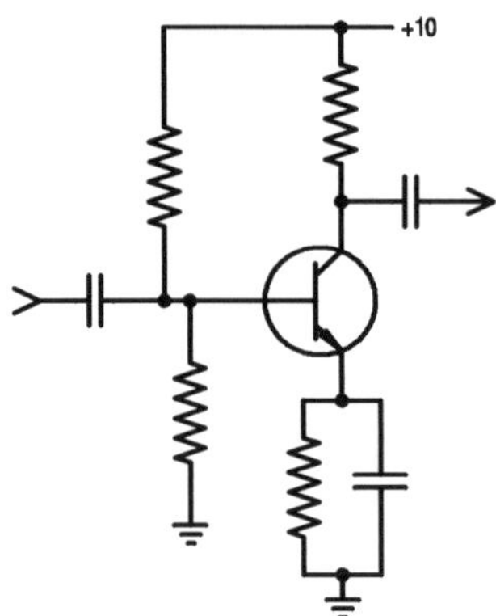

The first step is to choose a bias current for the transistor. Since no requirements were placed on us as to input or output impedance, we can be very flexible here! We will use 1mA to keep the calculations very simple. In order to give reasonable bias stability with temperature, we will choose to drop 1.5 volts across the emitter resistor. Ohms law will then immediately tell us that a 1K5 resistor is called for.

The collector voltage will swing upward and downward from its resting point. It cannot go any higher than the supply, and we want to keep it in the linear region, so we probably shouldn't let it get too close to the emitter voltage. Since we were not given any requirements for voltage gain, we will simply choose a collector load resistor that sets the quiescent collector voltage roughly half way between the supply voltage and about 1 volt above the emitter voltage. This is equal to (10+2.5)/2 = 6.25 V, which is 3.75 volts down from the supply. Therefore, Ohms law tells us that the collector resistor should be 3.75K. We will use the closest standard value, which is 3K9.

The AC input resistance to the transistor (r_π) is approximately 100 times 25/1, or 2500 ohms. We will use divider resistors that have at least 5 times this value so that the signal is not swamped. We will start off by defining the lower resistor (which we know will have the smaller value) at the standard value of 15K. The base voltage must be equal to V_{BE} plus the emitter voltage, or 0.7 + 1.5 = 2.2 V. The current going through the 15K resistor is therefore 2.2/15K = 0.147 mA. The upper base resistor must supply not only the 0.147 mA through the lower resistor, but also the base current, which is 1mA/100: the total current is therefore 0.157 mA. Ohms law is used to calculate the required resistance as (10-2.2)/0.157 = 49.68K. We will use the standard value of 47K, which will have the effect of slightly increasing the base voltage, and hence the quiescent collector current.

The input coupling capacitor must be of sufficient size that its reactance at the lowest frequency to be amplified is less than the parallel resistance of the two base bias resistors and r_π. The emitter bypass capacitor should be large enough that its reactance is much less than the emitter resistor at the lowest frequency to be amplified. Assuming that the lowest frequency to be amplified is 100Hz, the following circuit will be the result of our simple-minded design exercise:

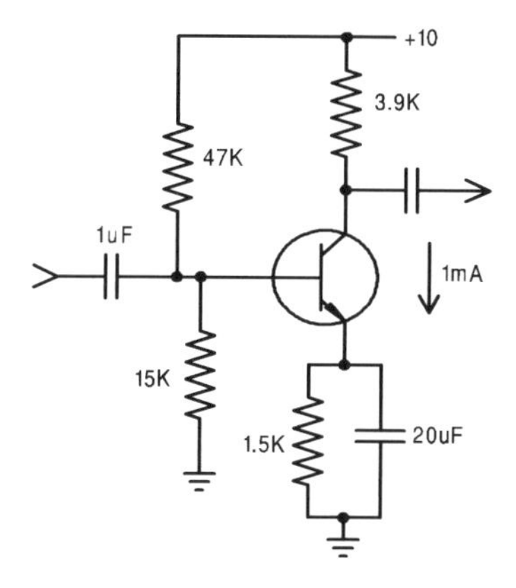

The voltage gain at low frequencies of this amplifier will be approximately:

$$A_V = R_L g_M = 3900 \frac{1}{25} = 156.$$

The output signal is 180 degrees out of phase with the input signal. Note that <u>if</u> the emitter resistor is <u>not</u> bypassed, several interesting things happen. The voltage gain becomes simply the ratio of the collector to the emitter resistors, or 2.6 in this case. It can be seen that the emitter bypass capacitor has a very large effect on the gain! If the emitter bypass capacitor is omitted, any signal on the base will cause a similar (but offset by V_{BE}) signal across the emitter resistor, and therefore a variation in collector current. However, note that from the Hybrid π model, the current generator is controlled by the <u>relative</u> voltage from the base to the emitter: any signal that occurs on the emitter will subtract from the relative input. This is a form of negative feedback: more will be discussed on this topic later.

The DC input impedance to the base becomes approximately equal to h_{FE} times the emitter resistor, or 150K. If the unbypassed emitter resistor is set equal to the collector resistor, and the bias resistors are set appropriately, then the circuit will have a gain of close to one, but two out-of-phase outputs will be available: one on the emitter, and one on the collector. This circuit is then called a "**phase splitter**", and it is useful in push-pull amplifier arrangements:

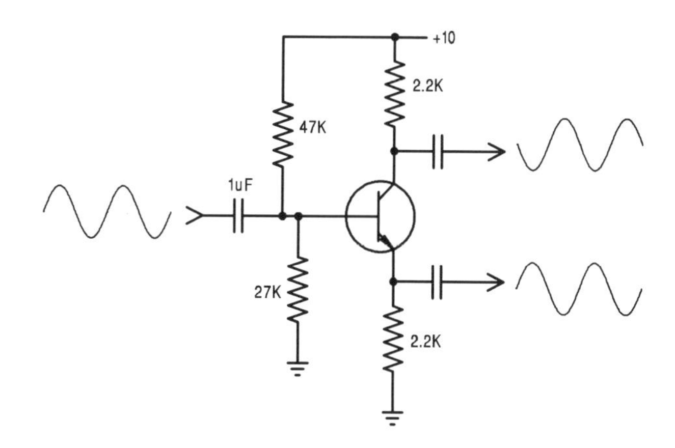

"Rules Of Thumb" for designing low frequency AC-coupled small signal transistor amplifiers

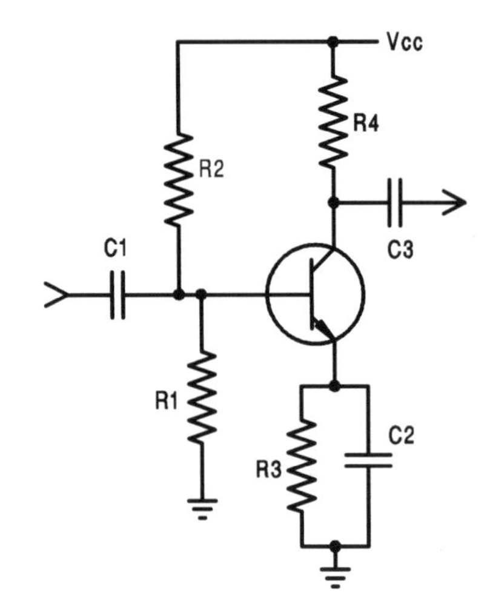

The following procedure works fine for frequencies of up to a few hundred KHz using general purpose transistors such as the 2N3904 (npn) or 2N3906 (pnp).

Note that V_{CC} will be positive if using an npn device, and negative if using a pnp transistor. The following procedure will work fine for values of V_{CC} which are greater than 5 volts.

1. Choose the operating (quiescent) current for the transistor. Unless there is a good reason to use a different value, use something between 0.5 mA and 2.0 mA. Call this value I_C.
2. Decide how much voltage to drop across the emitter resistor R_3. Remembering that the V_{BE} drop across the transistor is nominally 0.7 volts, and that it decreases by 2 mV per degree C, a good choice from a stability standpoint will be to drop about 10% of V_{CC} across R_3. Calculate R_3 using Ohms Law.
3. Choose the values of R_1 and R_2 to place a voltage on the base that is approximately 0.7 volts higher than that dropped across the emitter resistor. The parallel resistance of R_1 and R_2 should be higher than the AC input impedance of the transistor (described in step 5). For stability reasons, it is recommended that the value of R_1 be no more than about 20 times R_3.
4. Use Ohms Law to choose a value of collector load resistor R_4 such that the nominal voltage on the collector is approximately mid-way between V_{CC} and the voltage on the emitter.
5. The AC input impedance looking in to the base of the transistor is approximately equal to $(\beta+1)(R_U+25/I_C)$, where I_C is measured in mA, and R_U is the value of any unbypassed resistance in the emitter circuit. The DC input impedance looking into the base of the transistor is approximately equal to $(\beta+1)(R_E+25/I_C)$, where R_E is the total resistance (bypassed or not) in the emitter circuit. The amplifier input impedance is equal to the parallel combination or R_1, R_2, and the transistor base input impedance.
6. Choose C_1 to give the desired low frequency response. The low frequency response will be down by 3 dB at a frequency where the reactance of C_1 is equal to the input impedance of the amplifier as defined in step 5.
7. Choose the value of emitter bypass capacitor, C_2. We will start by assuming that the objective is to get the greatest possible voltage gain out of this amplifier. At the lowest frequency of interest, the reactance of C_1 should be no more than $25/I_C$, where I_C is measured in mA. In this case, all of the emitter resistor is bypassed for AC frequencies, and the voltage gain of the amplifier (assuming zero

source impedance and infinite load impedance) is approximately equal to (R_4 I_C)/25, where I_C is measured in mA. Better linearity (meaning lower distortion) and better stability and reproducibility can be achieved by only bypassing a portion (or perhaps none at all) of the emitter resistor. Then the overall voltage gain becomes approximately equal to R_4/R_U, where R_U is the value of the unbypassed portion of the emitter resistance.

8. The output impedance of the amplifier is approximately equal to the load resistor R_4.
9. The high frequency roll off of the amplifier depends on the load capacitance and on the transistor's transition frequency F_T. As a very rough approximation, the maximum voltage gain that can be achieved at a given frequency can be determined by dividing F_T by the frequency.

A "**darlington**" transistor is actually two transistors in one package, connected as shown below:

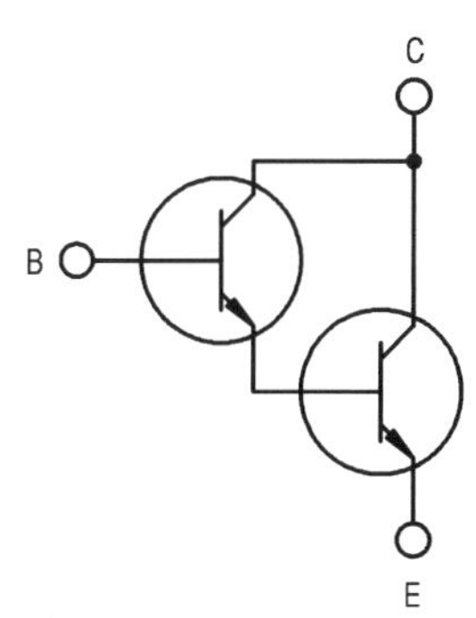

Although they are commonly available as a single package, a darlington configuration can easily be made by connecting two conventional transistors in the same arrangement. The h_{FE} of a darlington is equal to the product of the h_{FE}'s of the two individual transistors. The V_{BE} of a darlington is twice as high as for a single transistor.

An interesting variant on the darlington is the so-called "**complementary darlington**" configuration:

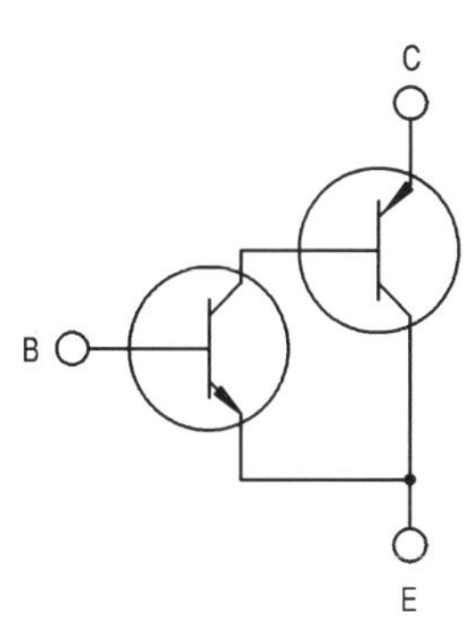

The complementary darlington makes use of one npn and one pnp transistor. The effective V_{BE} of the combination is approximately 0.65 volts.

Transistors are often used as switches, especially as part of TTL logic. When a transistor is turned "hard on" by injecting substantial base current, the collector voltage is pulled low, and the transistor goes into saturation. The collector voltage can go below that of the base (NPN transistor assumed). When it comes time to turn the transistor "off", there will be a measurable delay due to stored charge. If substantial collector current is being "sinked", it may be impossible to get V_{CE} to go below 2 volts or so.

High Frequency Performance of RC-Coupled Transistor Amplifiers

So far we have only discussed the low-frequency characteristics of a simple NPN transistor amplifier stage. At higher frequencies we need to take into account the equivalent input capacitor C_π. For very low frequencies, the reactance of C_π is much higher than the value of r_π, and its effect can be ignored. As the frequency increases however, C_π's reactance becomes lower than r_π's resistance, and gain is decreased.

If there are several stages of RC-coupled amplification, the effective load impedance of one stage is the parallel equivalent of the load resistor and the input impedance of the following stage, which will be dominated by C_π at higher frequencies. For the transistor amplifier described above, the frequency that this effect starts to be observed is when the capacitive reactance of C_π is equal to 2500 ohms. To analyze this effect, we must first determine the value of C_π as follows:

$$C_\pi = \frac{\beta}{2\pi f_T r_\pi} = \frac{100}{6.28 \cdot 2x10^8 \cdot 2500} = 3.18x10^{-11} = 31.8\,pF$$

The frequency at which the reactance of C_π is equal to r_π can then be calculated, yielding a value of 2 MHz. From this frequency upward, the gain will roll off at 20 dB per decade (6 dB per octave) for each stage. The actual roll off will be worse than this, because there is also capacitance between the collector and base to be considered. This C_{cb} is directly connecting the input terminal of the amplifier stage with a higher, out-of-phase signal at the output. The amplification of the stage increases the "effective" value of C_{cb}, further decreasing the high frequency performance.

Lead resistance, lead inductance, and stray circuit capacitance will all further degrade the high frequency performance.

Later chapters will provide design information on amplifiers working at higher RF frequencies.

POWER SUPPLIES

Almost all active circuits require a source of DC power. Some portable equipment is designed to operate directly from a battery, but most bench equipment is normally connected to a standard AC outlet which provides 120 VAC at 60 Hz in North America. European outlets provide 240 VAC at 50 Hz, and Japanese power is 100 VAC at 50 or 60 Hz. Some countries of the world supply 220 VAC at 50 Hz.

If a product is being designed with a conventional transformer type of power supply, it is possible to incorporate a switch in the transformer's primary circuit which will configure two separate tapped primaries in either series or parallel in order to accommodate the supply voltage available in any country in the world. Modern switching power supplies can be designed to connect directly to the line voltage, and can accommodate any input voltage from 95 VAC to 250 VAC without any adjustments.

We will start this discussion on power supplies by addressing the conventional transformer input linear power supply. A typical schematic for a simple power supply to deliver 12 VDC from a 120 VAC input is shown below:

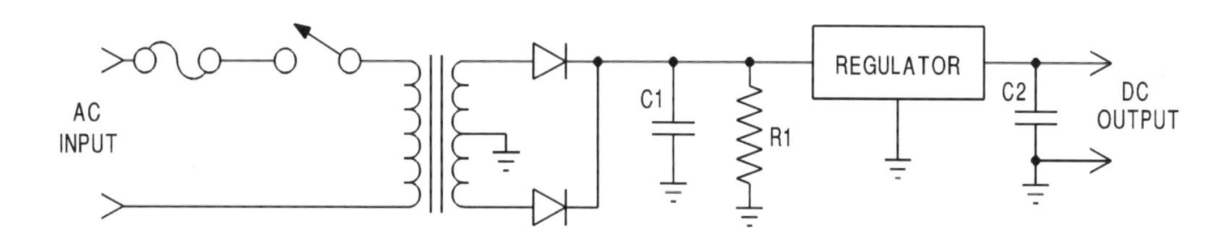

In this schematic, a centre-tapped transformer is used so that only two diodes are necessary to perform full-wave rectification. If a centre-tap is not used, four diodes (forming what is referred to as a full-wave bridge) are needed to get full wave rectification. The diodes can be conventional silicon devices such as the 1N4002, or they can be Schottky rectifier diodes that offer a lower forward voltage drop. The use of low drop diodes can slightly increase efficiency and decrease heat generation.

In the absence of C1, the input to the regulator would look similar to the left hand waveform shown below. Whenever the voltage dropped down below the regulator's reference voltage, the output of the power supply would also fall off, causing an output with an enormous amount of 120 Hz ripple. When C1 is included in the circuit, it maintains the input voltage to the regulator in between the charging pulses from the diodes. The more current that is drawn, the more the voltage across C1 will "droop" between cycles. More capacitance will decrease the amount of "droop", or ripple. The right hand waveforms below show the effect of smaller or larger values of C1 for the same output current.

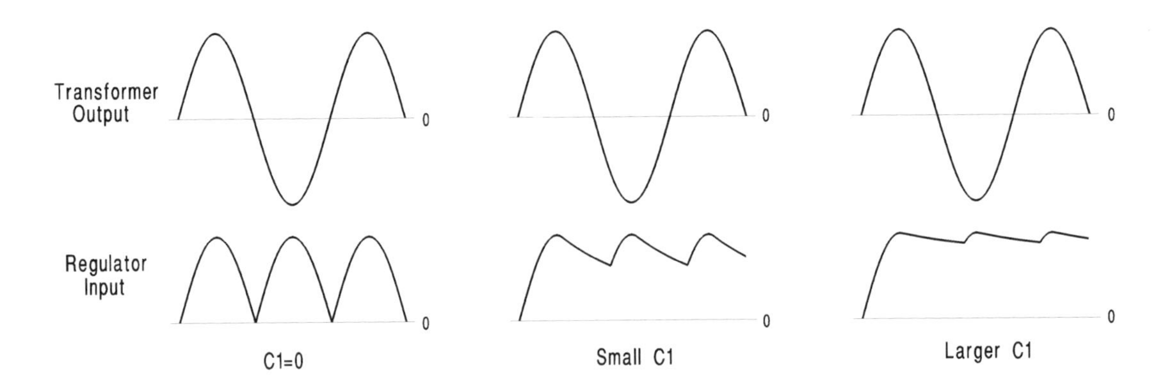

The bottom of the ripple must not fall below the minimum required input for the regulator. If one of the commercially available 3-terminal regulators (such as the 7812) is used, the "dropout voltage" (minimum allowable difference between the input and output voltage) will be about 2 volts, therefore the trough of the ripple must be at least 14 volts if the supply is designed for a 12 volt output. In sizing the capacitor C1 and the transformer specifications, it must be recognized that the AC input voltage from the line could be as low as 103 VAC under some circumstances.

The peak-to-peak ripple voltage is a function of the line frequency, the capacitance, and the load current. For a 60 Hz system, the peak-to-peak ripple can be calculated as $V_{ripple} = 8000\ I / C$, where I is the DC load current in amps, and C is measured in μF.

A smaller value of C1 can be compensated for by a larger number of turns on the secondary of the transformer - even though the ripple is high, the troughs still remain above the regulator's dropout voltage. Unfortunately, this will result in more power having to be dissipated in the regulator, and less overall efficiency.

The power dissipated in the regulator is equal to the load current times the difference between the regulator's input and output voltages. This calculation should be done under the "worst case" conditions of maximum expected load current, and maximum line voltage (about 132 VAC in North America). If the power exceeds about one half watt, some type of heat sinking will be required in order to prevent the regulator from overheating and shutting down or failing.

The capacitor marked as C2 is an important component to include. It does not need to be a very large value (1 μF or so is fine), but it should have a low internal impedance. Its purpose is to supply or sink current for very rapid load variations, and to prevent the regulator from oscillating. The regulator has a low output impedance for low frequency variations, and the capacitor ensures that there is also a low source impedance to higher frequency variations.

Resistor R1 is a "bleeder" resistor which serves to discharge the capacitor C1 when the power supply is turned off. If R1 is too large, it will take an excessive time to discharge C1. If R1 is too small, it will dissipate appreciable power.

Three terminal regulators (the terminals are labeled as "In", "Out", and Ground) are available from several manufacturers in a variety of voltages, polarities, and power capacities. These devices typically include thermal protection and current limiting. For higher power applications, commercial regulators are available that are designed to be used with an external power transistor that is properly heat sunk. Low dropout regulators (less than 1 volt) are available for low currents (up to about one quarter amp), but the usual dropout voltage is at least two volts. Packaged regulators are also available for variable output voltage applications (the devices have more than 3 terminals).

Rather than using commercial regulators, the die-hard fanatic can always build one from scratch. A stable reference voltage such as a zener diode is needed, as well as a power transistor to pass the current and some type of error amplifier to compare the output voltage with the desired output. The simplest regulator is just a zener reference and an emitter follower, two examples of which are shown below:

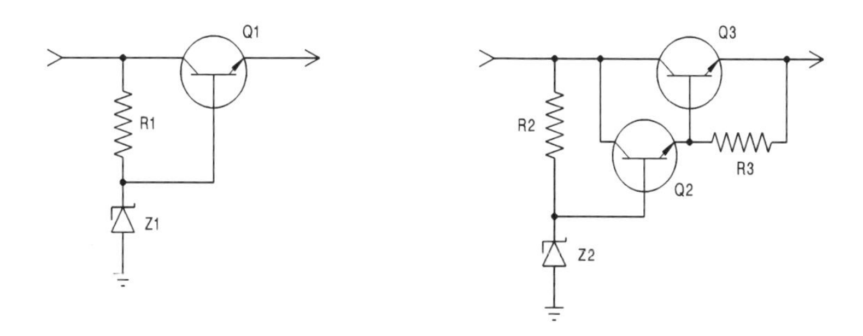

The left hand regulator is certainly simple, but its performance is rather anemic. The output voltage is nominally 0.7 volts less than the zener diode voltage, and it will increase as Q1 warms up (V_{BE} decreases by 2 mV per degree C). R1 must provide sufficient current to keep Z1 clamping hard, as well as the base current for Q1. The β for Q1 could be as low as 25 for currents of 100 mA or more. Consider an example design which was supposed to provide one half amp at 12 VDC. Assume that the input voltage minimum (bottom of the ripple's trough) is 15 volts at low line conditions. The base current for Q1 will be $500 / \beta = 20$ mA. Assume that the zener current should be at least 10 mA. This indicates that R1 will need to be $(15\text{-}12.7)/0.03 = 77$ ohms. Now consider the situation where there is no load, and the line voltage is high such that the average input voltage to the regulator is 20 VDC. The current through the zener will now be $(20\text{-}12.7) / 77 = 95$ mA, and the power dissipated in the zener will be $12.7 \times 0.095 = 1.2$ Watts. This will require that a 2 Watt zener be used instead of the smaller, cheaper, and more stable 250 mW device.

As the load current increases, Q1's V_{BE} will increase, and the zener's voltage will decrease (due to less available current). The net result is that the regulation will be poor: the output voltage will decrease somewhat as the load current increases.

The right hand circuit uses a darlington configuration to decrease the amount of base current that needs to be supplied by the reference (Z2 and R2). This will decrease the zener's dissipation, but will result in two other problems: there are two V_{BE}'s between the reference and the output, and the dropout voltage requirement will be higher because there are two transistors that need to be actively be biased. Resistor R3 is necessary to prevent thermal runaway if Q3 becomes hot enough that the I_{CBO} leakage current from Q3's collector to its base becomes sufficient to turn on Q3 even in the absence of any external base current.

Zener diodes are only available in certain voltages, and they can vary between units by 5% or more. The actual zener voltage will also vary somewhat with current and temperature.

The simple emitter follower regulators shown above have many disadvantages. A more complex but better solution is to use a circuit such as the following:

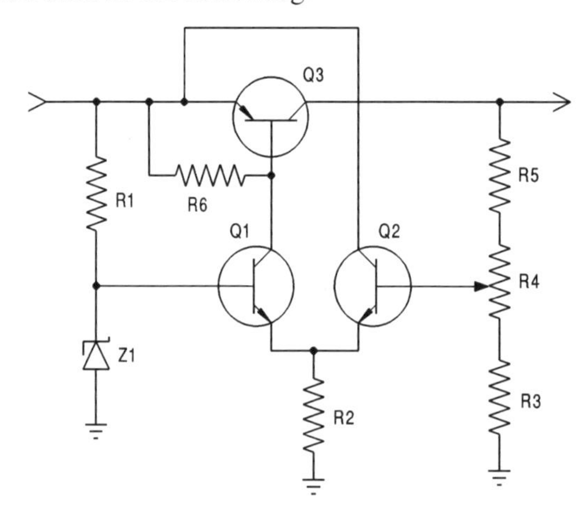

In this regulator, Q1 and Q2 form a "differential amplifier", which is a stable DC configuration that will be discussed in the next chapter. Any difference in the voltages of the bases of Q1 and Q2 will result in a change in the current through Q1, and hence the base current in Q3. Q3 is a pnp transistor which handles the load current: its V_{BE} variations do not effect the output voltage. The zener voltage Z1 is chosen to be approximately equal to one half of the output voltage: variations are taken care of by the adjustment at R4. R6 is again used to prevent thermal runaway of Q3. Because of the pnp arrangement of Q3, the dropout voltage of this regulator is fairly small. If for any reason (such as increased load current) the output voltage were to fall slightly, it would cause more current to flow through Q1, which would increase Q3's base current, causing the output voltage to increase, nullifying the drop. This is a form of negative feedback, and it reduces the regulator's effective output impedance.

Note than none of these discrete regulator designs incorporate current limiting, short circuit protection, or thermal protection. If these features are important, more circuitry needs to be added, or you can always resort to buying an off-the-shelf device.

Considerable power is dissipated in the form of heat for high current power supplies. The difference in Watts between the AC power going into the power supply to the DC power being supplied from its output is all lost in heat. This is a characteristic of linear regulators. Switching power supply designs were developed in the late 60's in an attempt to increase efficiency and decrease dissipated heat.

A switching regulator addresses the chief heat-producing component of a regulator: the pass transistor. An example will prove the point. If the input to a linear regulator is 12 VDC, and the output is 5 VDC, the pass transistor dissipates 14 Watts of heat if the load current is 2 amps. A switching regulator will use a pass transistor that is alternately either turned completely off (dissipating no heat), or hard "on" (where the power dissipation is perhaps 1 Watt). The transistor will be switched between the two states at a moderate frequency (perhaps 20 to 100 KHz) with a duty cycle of 41.7%. The average output will be 12 x 41.7% = 5.0 volts, and the power dissipated will be 1 x 41% = 410 mW. An LC low pass filter will remove all of the switching signal from the output, leaving only the DC component.

The heaviest and largest component in most power supplies is the transformer. So-called "off line switching supplies" (such as those used in most modern PC's) dispense with the heavy iron 60 Hz transformer, and instead use a diode bridge that is connected directly across the line voltage. After filtering, the capacitor holds over 200 VDC, and this is regulated down to the desired output voltages by means of a high frequency switching system using a very small transformer and variable duty cycle. If designed appropriately, this concept can be used to create a power supply that can operate from any line voltage in the world without any adjustments.

DC AMPLIFIERS

So far we have primarily been discussing AC amplifiers. There is often also the need to amplify small DC voltages. Because the signals to be handled have zero frequencies, it is obvious that coupling or bypass capacitors cannot be used.

Going to the example single transistor amplifier that we just discussed, it could be used to amplify DC voltages if we were willing to live with the fact that the quiescent DC output voltage is "offset" from the DC input voltage. The voltage gain would be equal to the ratio of the collector and emitter resistances. Another problem would be caused by the instability of this circuit with temperature.

In order to compensate for the DC offset through a single amplifier stage, it is conceivable that a zener diode or similar device could be used to compensate. Alternatively, a mixture of NPN and PNP devices could be used in a multi-stage design. If the supply voltages were both positive and negative with respect to signal ground, the circuit could amplify small DC signals with minimal offset. Examples are shown below:

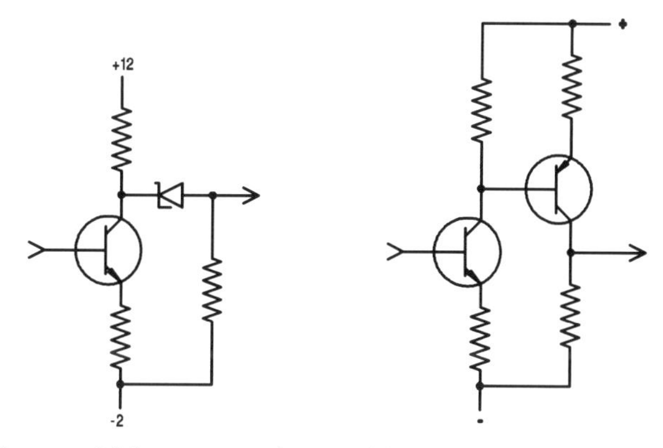

Unfortunately, these circuits would be extremely sensitive to variations in temperature, supply voltage, and batch variations. A circuit arrangement that has much better stability is the "differential amplifier", as shown below:

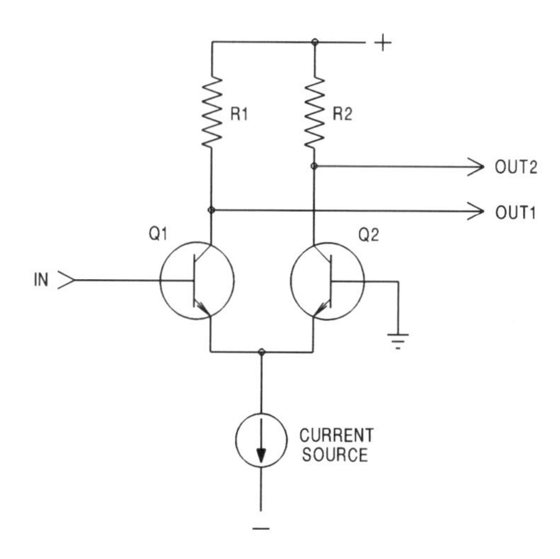

Looking just at Q1 for a moment, assume that it has an input signal of zero volts applied to its base. If both transistors are similar, and are at the same temperature (perhaps they are actually fabricated in the same package), the V_{BE}'s will be closely matched, and the two collector currents will be approximately equal. Notice that the sum of the two collector currents must equal the current flowing through the constant current source connected in the emitter circuit: as one transistor draws more current, the other must draw less.

The outputs are both positive with respect to ground. Two outputs can be produced: one at each collector. They will be of equal amplitude (with respect to the quiescent voltage), but opposite phase. The magnitude of either output is proportional to the difference in voltage between the two transistor bases. The absolute magnitude of the base voltages is not important, what is important is the difference between them. As temperature variations cause V_{BE} of one transistor to vary, the V_{BE} of the other transistor will follow suit, and the effect will be "common-moded" out. Differential circuits are used in almost all low level DC amplifiers.

Note that in this example, the base of Q2 is grounded. If instead of being at zero volts, the base is brought out to be another input, then the output of the differential amplifier will be a function of the difference between the two inputs. If a low level signal is to be connected to this amplifier via long cables, it makes sense to take both inputs of the differential amplifier out to the signal producing source (probably using a twisted wire pair). Any noise picked up by one wire will probably be picked up equally well by the other wire, and the net difference between the base signals on Q1 and Q2 will be zero. The configuration therefore is useful in reducing unwanted electrical noise pick up. The ability of a differential amplifier to reject common-mode signals is quantified in a term known as the "common-mode rejection ratio", and this is usually specified in dB. The current source can be a simple circuit such as the following:

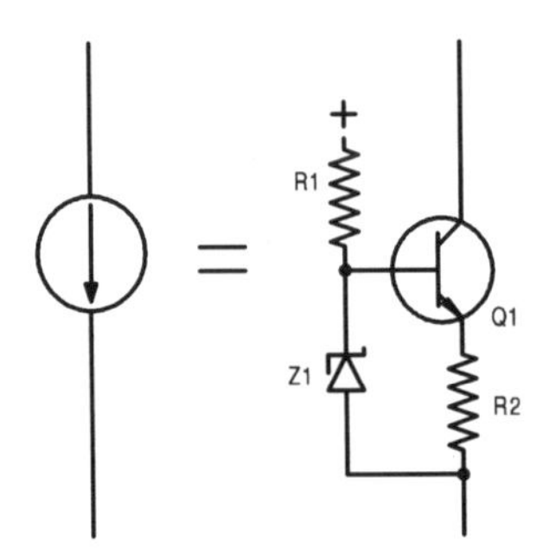

The zener diode Z1 is reverse biased from some positive supply by R1. This establishes a fixed voltage between the base of Q1 and the bottom of R2. Q1 functions as an emitter follower, and maintains the zener voltage (less one V_{BE}) across R2. Therefore, the current through R2 will be constant. The collector current is basically the same as R2's current (except for the small amount of base current), so the entire circuit acts as a constant current source (or sink).

Note that the single differential amplifier stage shown above still suffers from the problem that the outputs are offset (from a DC standpoint) from the input voltage. In order to solve this problem (and also to provide some more gain), it is common to follow one differential pair with another differential pair using the opposite polarity transistors:

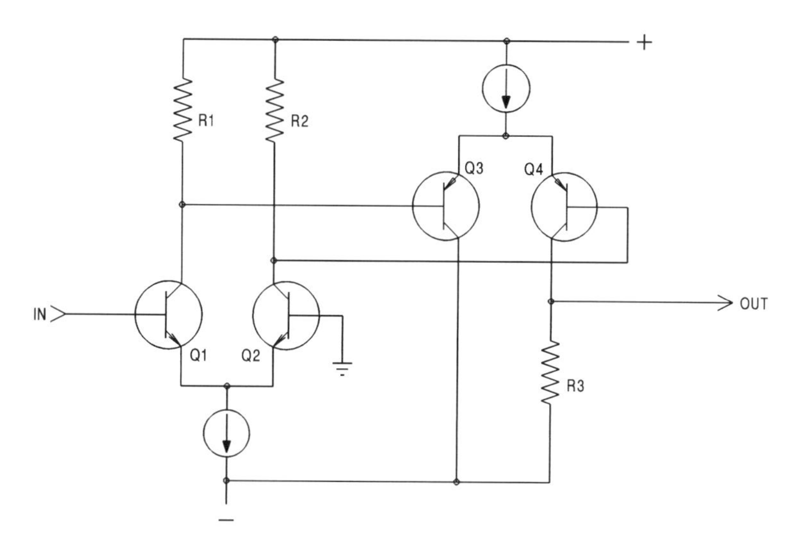

This circuit is drawn with only one input and one output. If differential inputs are required, the base of Q2 can be used as the second input. Similarly, a collector resistor can be added to Q3 to create a complementary output. Differential amplifiers can be used with FETs and tubes also. Operational amplifiers (discussed in the next section) use differential input amplifiers based either on bipolar or FET devices. A good operational amplifier will have an initial offset voltage at the input of only a few millivolts, and this offset will vary no more than a few microvolts per degree of temperature change. Here is a circuit of a <u>very</u> basic operational amplifier using these techniques:

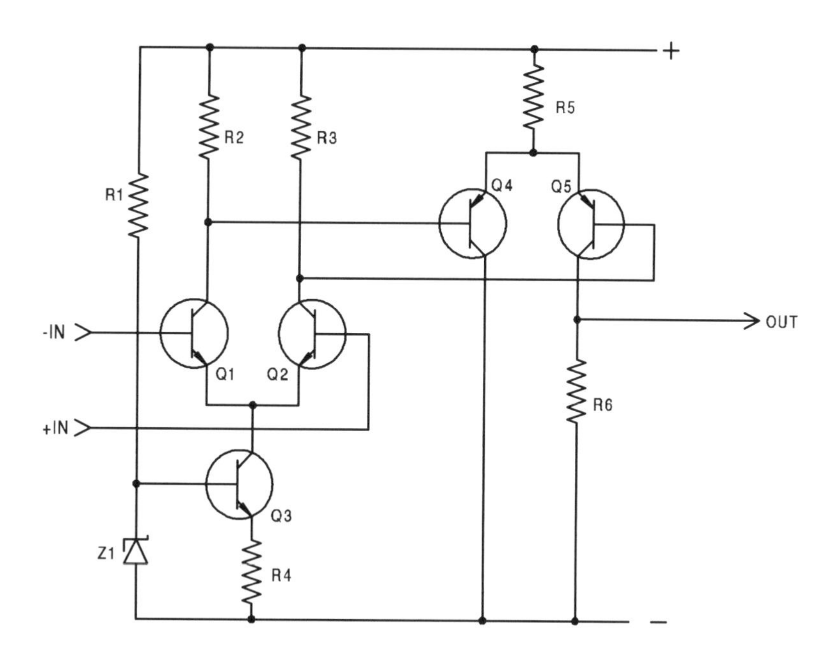

Note that the first differential amplifier has a proper current source, but the second stage just uses a simple common emitter resistor to approximate one.

OPERATIONAL AMPLIFIERS

The operational amplifier (abbreviated as "op amp") is an essential tool in the low frequency circuit designer's toolbox. Ever since the first monolithic 702 and 709 op amps were designed by Bob Widlar in the mid 60's, this component has become the building block of ever more complex analog systems.

The easy way to think of an op amp is to consider that it is a DC-coupled amplifier with closely matched differential inputs, a single output, high input impedance, and extremely high voltage gain (often over 100,000). The device is usually connected to both a positive power supply and a negative supply (plus and minus 12 or 15 volts are common) so that it can work with a range of signals that are on either side of ground potential. The schematic symbol for an op amp is:

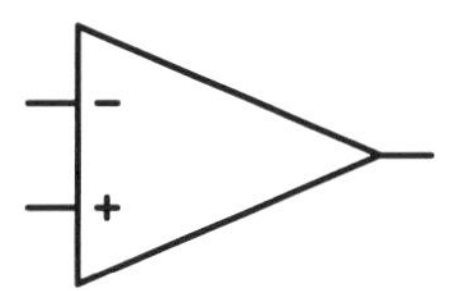

The "-" input is called the "inverting input": a positive-going signal on this pin will produce a negative-going output. The "+" input is called the "non-inverting" input: the output of the op amp is in phase with any inputs on this pin.

The op amp is hardly ever used in an open loop fashion: there are almost always components connected around it. The simplest configuration is the following basic amplifier:

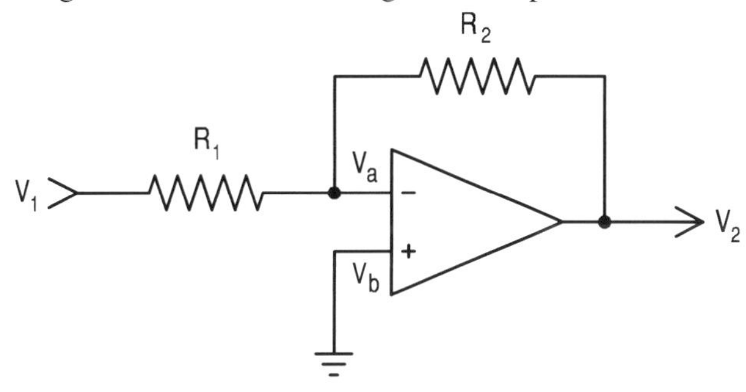

In order to understand how op amp circuits work, consider that the device has an extremely high gain (usually over 100,000), and that the output terminal will do anything that it has to do in order to make the two input terminals have an equal voltage. The input impedance is very large, and can often be ignored.

To analyze this simple circuit in more detail, consider that the op amp has a gain of G, and that G is quite large. The output voltage is therefore equal to $-G(v_a-v_b)$. We have drawn the circuit so that v_b is zero. Let's look at the currents going in to the inverting input pin: since the input impedance is very high, the sum of these currents must be zero. Therefore,

$$0 = \frac{V_1 - v_a}{R_1} + \frac{V_2 - v_a}{R_2}$$

Now, we know that v_a is equal to $-V_2/G$, so we can substitute as follows:

$$0 = \frac{V_1 - V_2\left(\frac{1}{-G}\right)}{R_1} + \frac{V_2\left(1 - \frac{1}{-G}\right)}{R_2}$$

However, G is very much greater than 1, therefore the term (1/G) approaches zero and can be ignored. Re-arranging the terms we are left with the expression for the gain of this circuit:

$$\frac{V_2}{V_1} = \frac{-R_2}{R_1}$$

The important thing to recognize here is that the gain of this circuit is only dependent on the ratio of the two resistors; the actual gain of the raw op amp has virtually no effect! This clearly represents a circuit that will be stable and easy to reproduce. The resistor from the op amp's output back to the inverting input is called the "feedback resistor". In this case, the feedback is negative. The voltage at the inverting input in this circuit is effectively zero. Remember that we are going to consider the op amp's gain to be very large, and therefore the output will swing whatever way it must in order to make sure that the two input pins have essentially the same voltage. Since the inverting pin is at very close to ground potential, the input impedance of this circuit is equal to R_1.

In applications where small signals are involved, and the demands for DC stability are stringent, it is customary to place a resistor in series with the non-inverting input. This resistor is chosen to have a value equal to the parallel combination of R_1 and R_2. The reason for this is that the input transistors of the op amp do draw a small amount of current (nano amps or pico amps), and this current will vary as a function of temperature. The current in each base will change by approximately the same amount for a given temperature change, so by keeping the DC resistances on the two inputs equal, these changes will common-mode out.

Now examine the following circuit, which shows the basic arrangement for a non-inverting amplifier using an op amp.

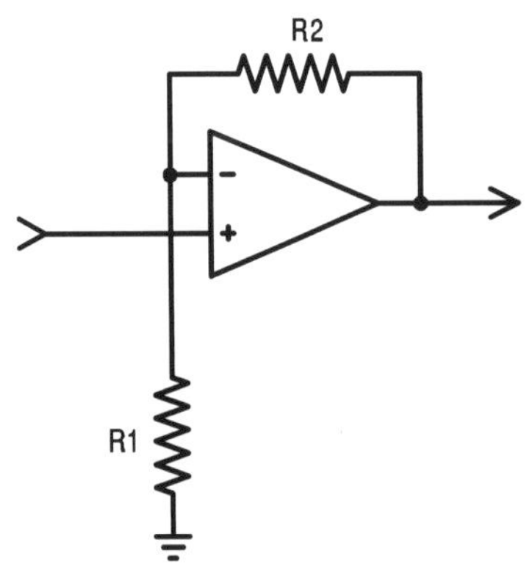

The voltage gain of this circuit is equal to $(R_2+R_1)/R_1$. and the input impedance is extremely high.

Both of these amplifier circuits have extremely small (approaching zero) output impedance. All of the desirable features of these op amp circuits are due to the fact that the open loop gain of the raw op amp is much higher than the closed loop configuration. The gain of the overall circuit has been reduced (and stabilized) through the use of heavy negative feedback. If the input and feedback resistors are adjusted so that the closed loop gain approaches the open loop gain, all of the easy-to-understand relationships break down. The op amp itself is made up of multiple bipolar and/or FET stages, and therefore its gain starts to roll off at some frequency. The product of the open loop op amp's gain at a given frequency and value of the frequency is a constant value, called naturally enough the "gain bandwidth product". This is specified in MHz. The open loop gain is equal to unity when the frequency is equal to the gain bandwidth product.

Rather than using two resistors, the input and feedback paths of an op amp configuration can use a variety of different components or networks. Some examples are shown below:

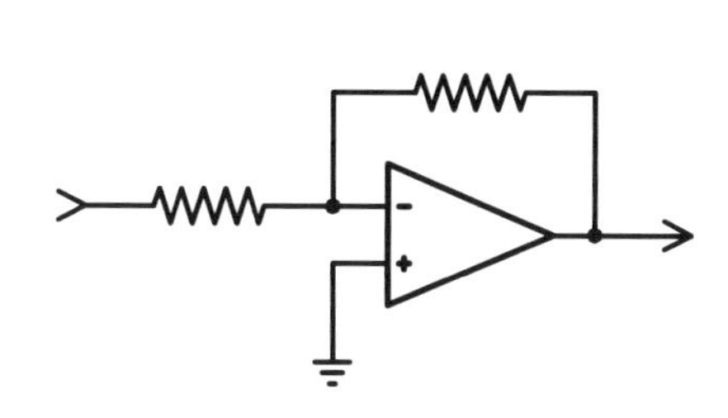

Basic Inverting Amplifier

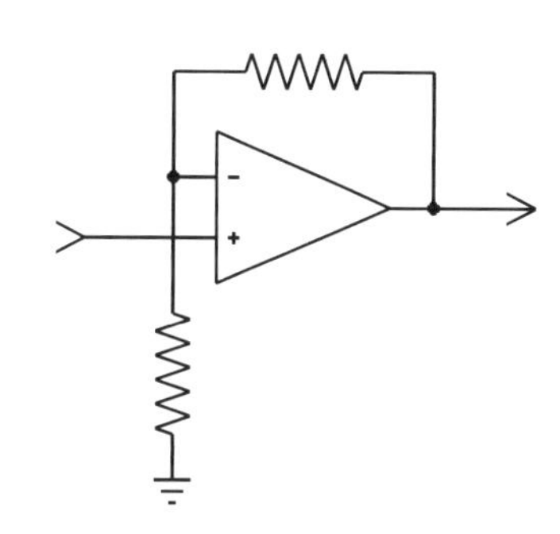

Basic Non-Inverting Amplifier

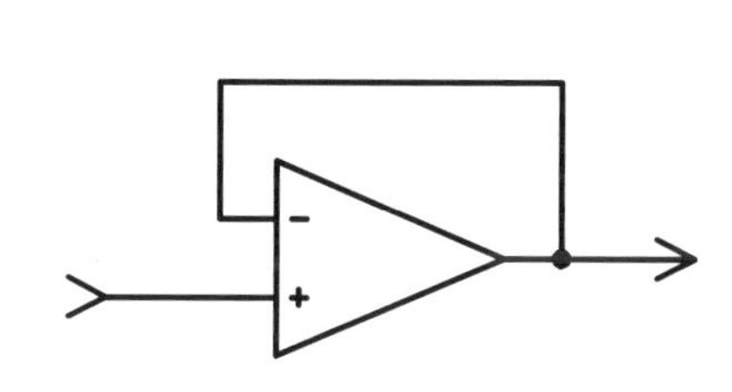

Unity Gain Buffer

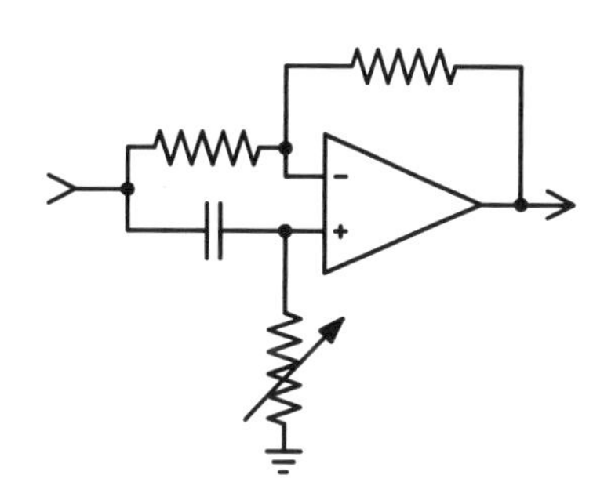

Phase Shifter

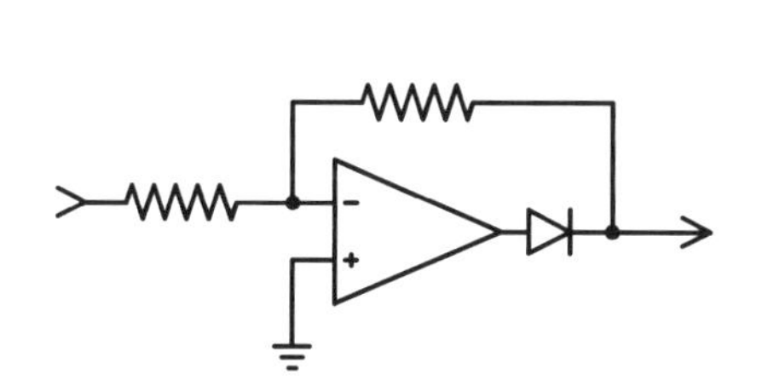

Half Wave Precision Rectifier

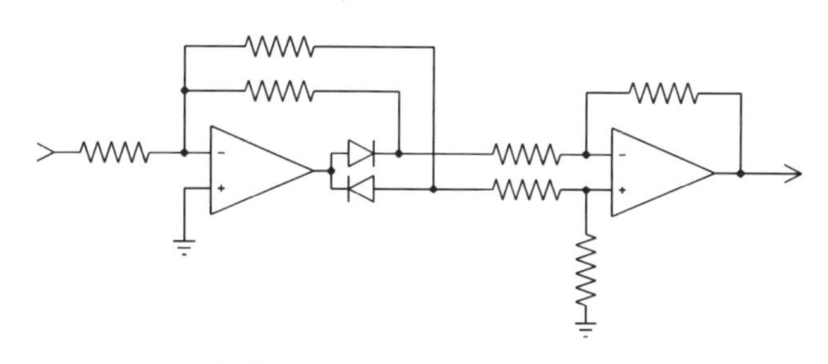

Full Wave Precision Rectifier

Note that the term "precision" means that the circuit rectifies the AC signal as though there was no forward voltage drop through the diodes.

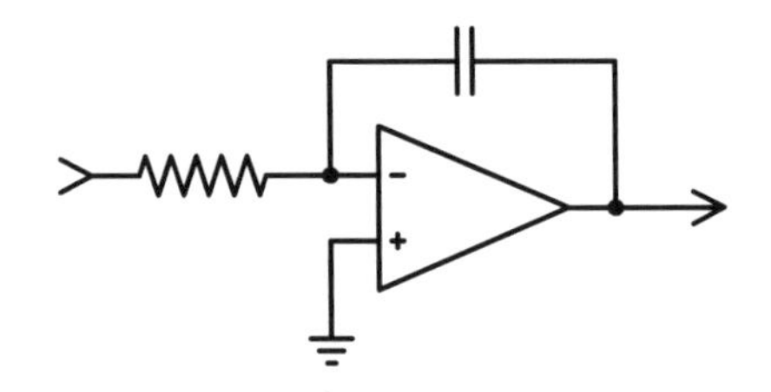

Integrator

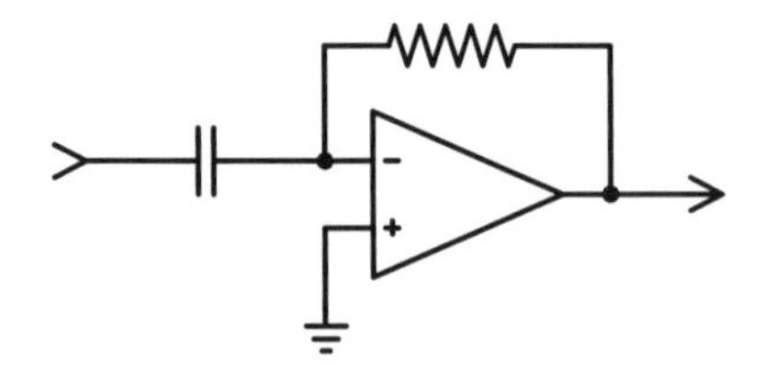

Differentiator

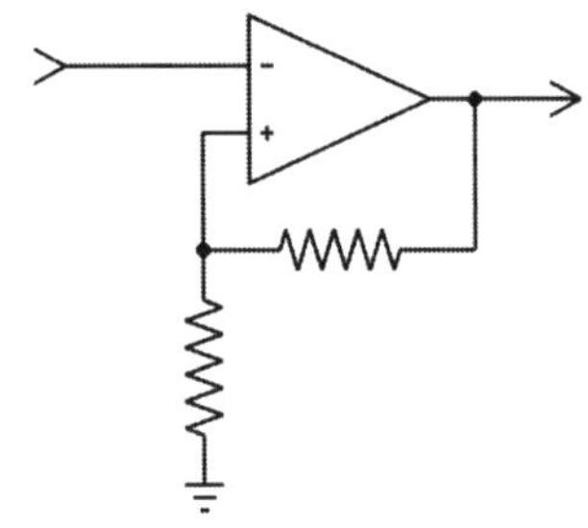

Comparator (with hysteresis)

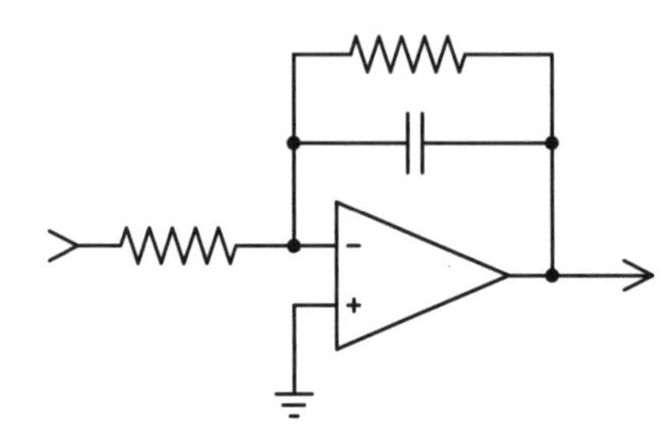

Low Pass Filter (single pole)

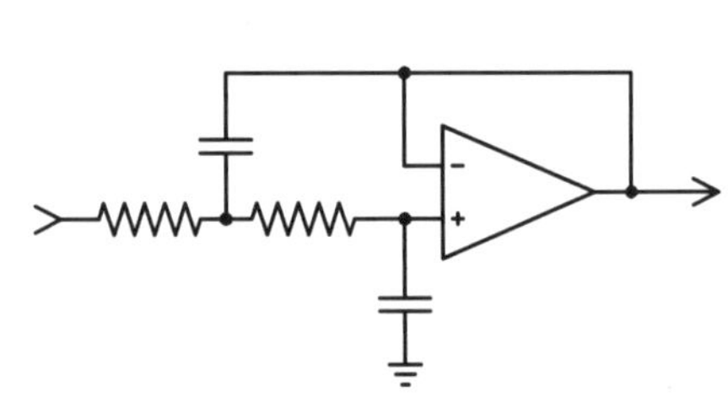

Low Pass Filter (two poles)

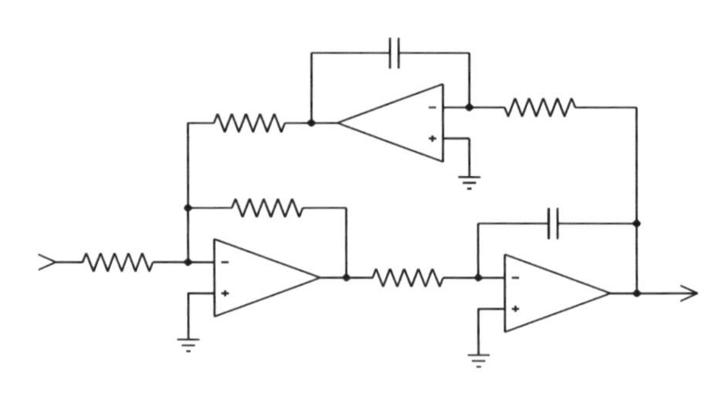

Band Pass Filter (State Variable Filter)

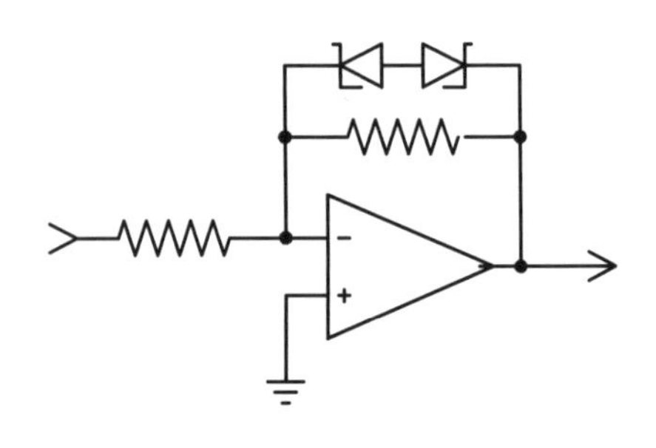

Clipper

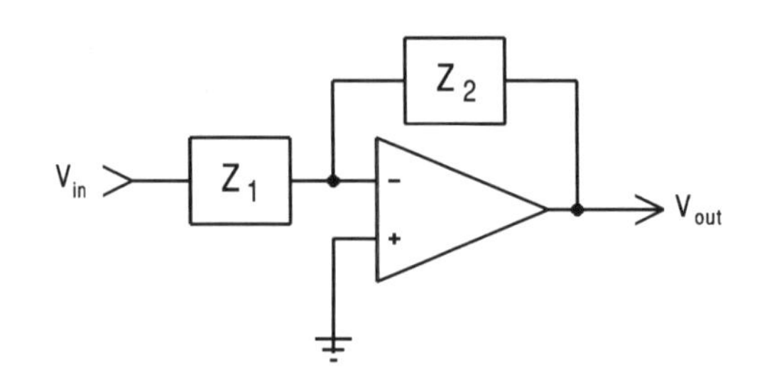

Generalized Op Amp Circuit: $\frac{V_{out}}{V_{in}} = \frac{Z_2}{Z_1}$

COMPLEX IMPEDANCES

Up to now, we haven't discussed the phase relationship of current and voltage in various electronic components. This concept is often not well understood, and it is important to grasp these ideas if you want to do any serious design work or circuit analysis.

The ARRL's "Advanced Class License Manual" give a good overview of some of the concepts, but without using imaginary numbers. This might be a good secondary source of information if you have difficulty with the following materials. The ARRL's "Extra Class License Manual" treats the topic properly, and introduces imaginary numbers. A review of these references would possibly aid in understanding the concepts which will be discussed in the following notes.

As we all learned during basic training, the current in a capacitor connected to an AC source leads the voltage by 90 degrees. The current through an inductor lags the voltage by 90 degrees. Because of this relationship, we must be especially careful in analyzing circuits that have both resistance and reactance. In order to better picture the phase relationship between the voltage across a component and the current through it, it is sometimes convenient to draw a polar diagram. The figure below is used for showing values that have both a magnitude and an associated angle. Positive angles are measured in a counter-clockwise direction, starting at the horizontal line which extends to the right. Negative angles are measured in a clockwise direction.

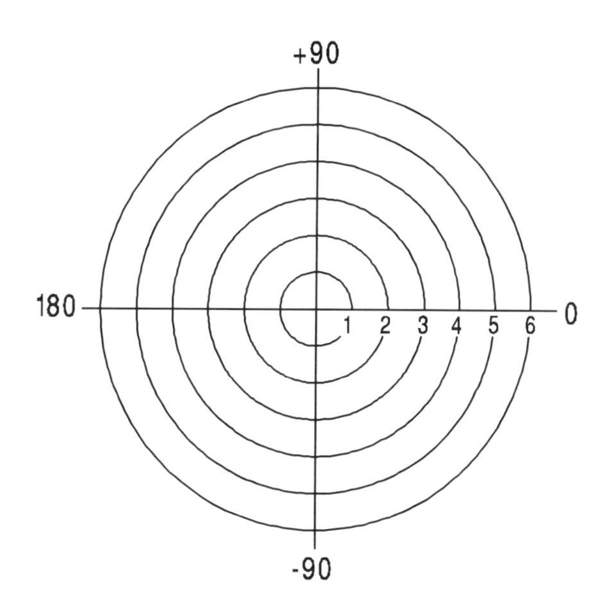

The circles are used to draw the magnitude values of the current or voltage. An arrow is drawn from the centre of the diagram out to the circle representing the magnitude in a direction representing the phase. As an example, the following three diagrams show the voltage across a single inductor, resistor, or capacitor as a 1 amp current at zero degrees flows through it. Note that the current is assumed to lie on the horizontal axis (at zero degrees): what is being plotted is the voltage across the component:

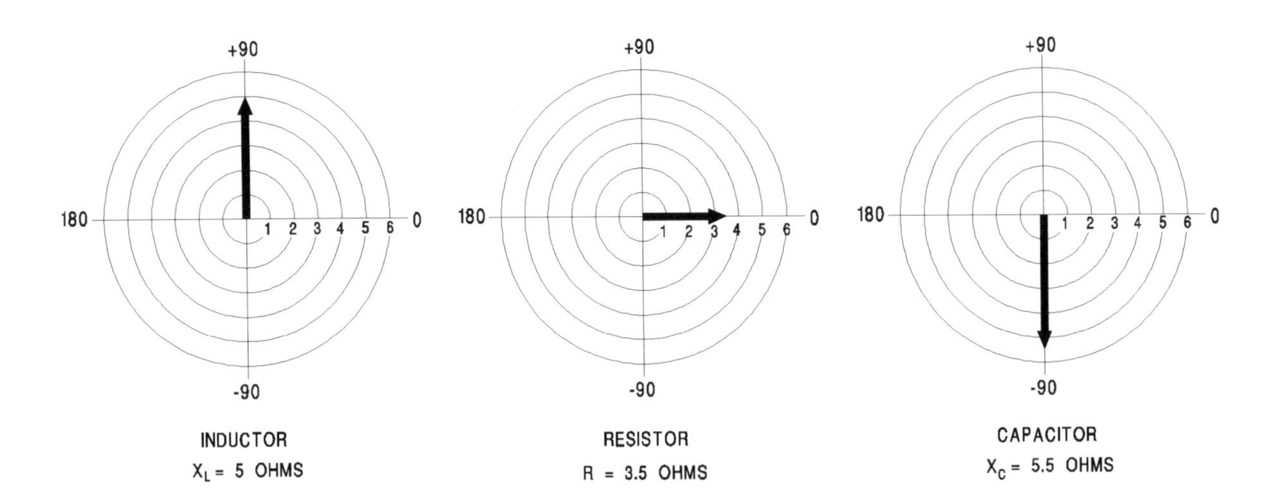

If a 100 ohm resistor is connected in series with a capacitor having a reactance of 100 ohms at the frequency of interest, the total impedance is **not** 200 ohms! Because the I-V phase relationship is different for resistors and capacitors or inductors, you can't simply add reactance and resistance values: this is like adding apples and oranges. In order to work with circuits containing reactive components, we need to use "complex impedance" methods.

The term "**complex**" does not refer to how difficult it is to understand the concepts, but rather to the type of numbers that are used. You may recall from high school that there are "real" numbers, such as 1, 2, 3, etc., and there are "**imaginary**" numbers such as $\sqrt{-1}$, $\sqrt{-2}$, etc. A complex number is one which contains **both** real and imaginary parts. An example of a complex number might be $12.5 + \sqrt{-3.6}$.

The square root of any number is that number which, when multiplied by itself is equal to the original number. In the real world, there are no two identical numbers which can be multiplied together to get a negative number. Therefore, the square root of any negative number is "imaginary". The square root of -1 is commonly used in many branches of engineering and the sciences. In all fields except for electronics, $\sqrt{-1}$ is called "**i**". Because this could be confused with the symbol for current, electrical and electronic texts refer to this special number as **"j".**

We will use "**j**" to represent a 90 degree phase relationship in the following discussion. Remember that whenever you see a number next to the letter "j", this refers to the magnitude that is at 90 degrees to the "in phase" component. As an example, if a 2 amp AC current flows through a 1 ohm resistor, the voltage is in phase with the current, and its value is 2 amps. If however a 2 amp AC current flows through a capacitor whose reactance is 1 ohm, the voltage will be 90 degrees out of phase with the current. This is written as -2j volts: the negative sign indicates that the voltage lags the current.

Remember that impedance is composed of two parts: resistance and reactance. The current and voltage are in phase for the resistive part of the impedance, but they are 90 degrees out of phase for the reactive part of the impedance. Another way of saying this is that Z = R + jX. Impedance is simply the ratio of the voltage across a component to the current passing through it. Since the current and the voltage will not necessarily be in phase if the component has reactive elements, we must express the impedance using a form that indicates not only the magnitude of the impedance, but also the phase relationship (in degrees) between the voltage and the current.

In this text from this point on, we will recognize that any expression of impedance has not only a magnitude, but also an associated angle. The angle just indicates the phase relationship between the voltage across the impedance compared to the current flowing through it.

Now, let's examine a simple circuit containing a 100 ohm resistor in series with an inductor whose reactance is equal to 100 ohms at the frequency of interest. We would like to determine the impedance of the total network at that frequency. Remember that current through an inductor lags the applied voltage by 90 degrees. This is another way of saying that the voltage across an inductor leads the current flowing through it. Since the resistor and the inductor are in series, the **current** through both components must be the **same**. The voltage across the inductor will appear to be 90 degrees ahead of the voltage across the resistor, and the voltage across the entire network (with respect to the current) will be the sum of the two voltages, **including** the effect of their phase differences. This can best be explained graphically. We will assume that a current of 1 amp flows through the circuit at zero degrees, then plot the voltages across the two series components:

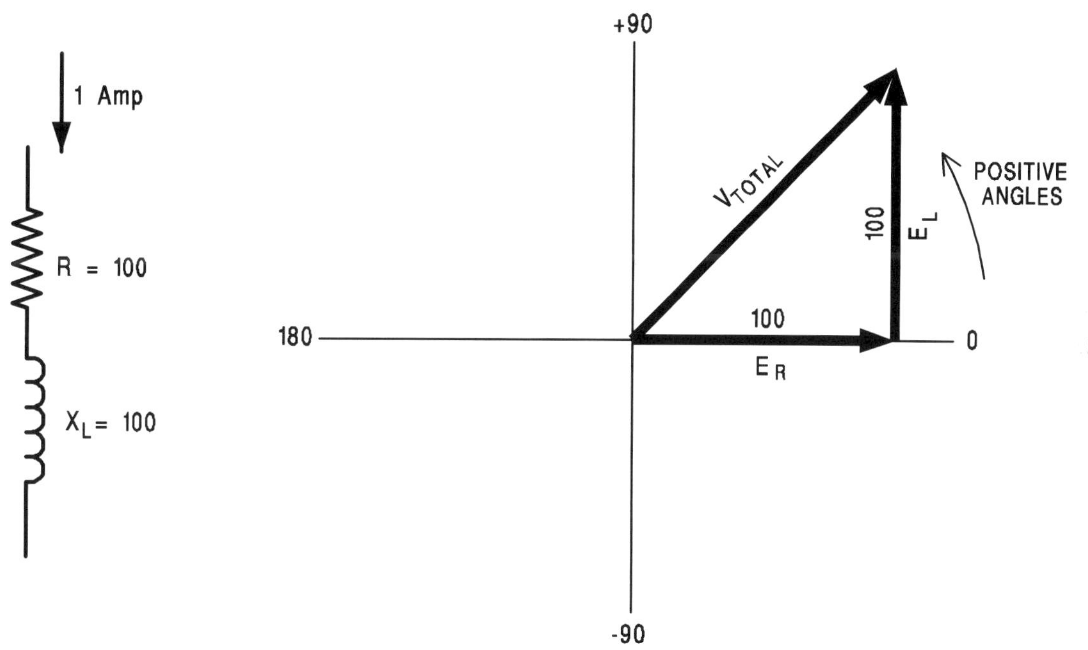

In this diagram, positive phase is measured (by convention) in a CCW direction, starting at 0, which is represented by the right hand horizontal axis. The circles have been omitted for clarity. The voltage across the resistor is exactly in phase with the current going through it, so 100 volts would be developed across it at an angle of 0 degrees. This is shown by the **vector** pointing to the right. Note that the term "vector" is a fancy way of saying something has both a magnitude and an angular relationship. Since the end of the vector represents the voltage across the resistor, we can now start from the end of the vector (where the arrow head is), and add the vector representing the voltage across the inductor. Since we know that the voltage across the inductor is 90 degrees ahead of the current, we will draw it at that angle.

The total voltage across the circuit is the distance from the origin (where we started from) to the end of the vector representing the voltage across the inductor. Note that this total voltage is itself a vector, having both length and direction.

The horizontal axis is referred to as the "**real**" axis: the voltage and current are in phase, so real power is dissipated. The vertical axis is referred to as the "**imaginary**" axis, where there is a 90 degree relationship.

The total voltage in this circuit can be determined by measuring the length of the vector on the graph, or it can be found by using simple geometry. Since this is a right angle triangle, we know that the length of the hypotenuse is equal to the square root of the sum of the squares of the lengths of the other two sides. (*This is the so-called "Pythagorean Theorem"*). Therefore the total voltage can be determined as:

$$V_{Total} = \sqrt{100^2 + 100^2} = 144 \text{ volts}$$

The phase angle for the applied voltage with respect to the current can be measured from our graph, or can be calculated using simple trigonometry. Since this is a right angle triangle, the angle θ is simply arctangent of the ratio of the two 100 volt sides of the triangle, or 45 degrees. At this point it might be a good idea to review the introduction to trigonometry that is outlined in the Appendix.

> **Note:** arctangents can be found by pressing the $\tan^{-1}$ button on calculators, or by looking up in trig tables. Some common and easy to remember values are:
>
> $\tan^{-1}(0)$ = 0 degrees
> $\tan^{-1}(1)$ = 45 degrees
> $\tan^{-1}(\infty)$ = 90 degrees

We now know that if a 1 amp AC current is flowing through this circuit, the voltage across it will be 144 V, and it will lead the current by 45 degrees.

The impedance of the circuit is simply the voltage divided by the current, which is therefore 144 ohms with a phase angle of +45 degrees. Note that impedances have both magnitude and phase!

The impedance of this circuit can be expressed as $144\Omega \angle 45$. This is the polar representation of the impedance: it tells you that the vector for the impedance is 144 units long, and is at an angle which is 45 degrees CCW from the 0 degree axis.

A more common way for expressing impedances is in terms of Cartesian coordinates, which simply identify the end of the vector in terms of the horizontal and vertical coordinates. The horizontal axis is the so-called real axis and the vertical axis is referred to as the imaginary axis, which is measured in "**j**" units. Our total impedance in this simple example is therefore: 100 + j100 ohms. Rather than refer to "Cartesian" coordinates, we will in the future refer to the more common term, which is **rectangular** coordinates.

We will now work out another example. Consider a circuit containing a 25 ohm resistor in series with a capacitor having a 50 ohm reactance at the frequency of interest. Again we will assume that a current of 1 amp is flowing through the series circuit. We will plot the voltage vectors as follows:

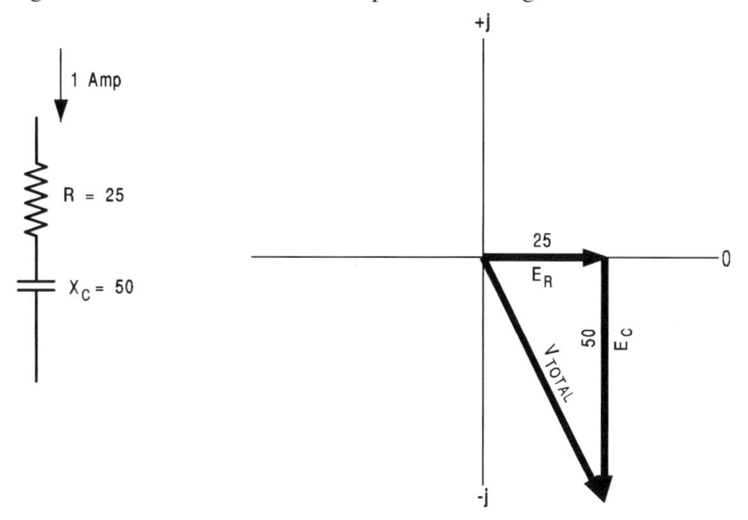

Note that the voltage across a capacitor lags the current, so it is drawn with a negative 90 degree angle. Using rectangular coordinates, we can directly see that the coordinates representing the total voltage are 25 along the real axis, and minus 50 on the imaginary axis. We can therefore say that the total voltage across this circuit is 25 - j50 volts. Dividing by 1 amp to get the total circuit impedance, the result is 25 - j50 ohms. See how easy this is!

Now, if you wish to convert this value to polar coordinates, use simple geometry. The magnitude can be determined using the Pythagorean Theorem, and the phase angle can be found by using the arctangent:

$$\text{Magnitude} = \sqrt{25^2 + 50^2} = \sqrt{625 + 2500} = 55.9 \text{ ohms}$$

$$\text{Angle} = \tan^{-1}(-50/25) = \tan^{-1}(-2) = -63.4 \text{ degrees}$$

Therefore, this impedance can be expressed either as 25 - j50 ohms, **or** as 55.9 ohms $\angle -63.4°$.

In general, impedances are expressed as R + jX, where R is the value of the resistance in ohms, and X is the value of the series reactance at the frequency of interest. It is possible to convert between the two representations as follows:

Rectangular to Polar - $\text{Magnitude} = \sqrt{R^2 + X^2}$

$$\text{Angle} = \tan^{-1}\left(\frac{X}{R}\right)$$

Polar to Rectangular - R = Magnitude x cos(angle)
X = Magnitude x sin(angle)

Based on our new understanding of complex impedances, we can now redefine the formulas for reactance.

$$X_C = \frac{-j}{2\pi f C} \quad \text{or} \quad X_C = \frac{-j}{\omega C} \text{ using angular frequency notation.}$$

$$X_L = 2\pi f jL \quad \text{or} \quad X_L = j\omega L \text{ using angular frequency notation.}$$

Now we will look at an example of a series circuit containing all three types of passive circuit component: a resistor, capacitor, and inductor. Let's consider a 200 ohm resistor in series with a 1000 pF capacitor and a 100 μH inductor. We would like to know what the total impedance is at a frequency of 1 MHz. First we have to calculate the reactance of the capacitor and the inductance at 1 MHz. Here, X_C = -j 159 ohms, and X_L = j 628 ohms. Because all of the components are in series, we can add their complex impedances as follows:

$$Z_{Total} = (200 + j0) + (0 - j159) + (0 + j628) = 200 + j469 \text{ ohms.}$$

Note that the real and imaginary components are separately added algebraically (*this means that we keep track of the signs*). Since the end result of this addition has a positive j component, we know that overall this circuit is somewhat inductive.

RULES FOR PERFORMING MATH ON COMPLEX VALUES:	
Adding or Subtracting:	Using the rectangular forms, add or subtract the real and imaginary terms separately.
Multiplying:	Using the polar form, multiply the magnitudes and add the angles.
Dividing:	Using the polar forms, divide the magnitudes and subtract the angles.

Now, let's calculate the series impedance of this same circuit at a frequency of 503 KHz. The reactances are: X_C = -j316 ohms, X_L = j316 ohms. If we now add up the impedances of all three components, we find that the two imaginary components cancel each other out, and all we are left with is a pure resistance (*no imaginary component*) of 200 ohms. This is **series resonance**! If the frequency were lower, the capacitive reactance would be higher, and the inductive reactance would be lower: the circuit would then have a negative j component, and the circuit would be somewhat capacitive.

OK, on to parallel circuits!

The current is not necessarily the same in all the legs of a parallel circuit, but the voltages are the same. Rather than talking about impedances, we will talk about the **"dual"** of impedance, which is called **admittance**. It is easy to think of "admittance" as being the reciprocal of "impedance". A circuit having more admittance will in general allow more current to flow, whereas a circuit having more impedance will allow less current to flow. The dual (or reciprocal) of resistance is **conductance**; the dual (or reciprocal) of reactance is **susceptance**.

Admittance, conductance, and susceptance are measured in **mhos** (or **Siemens**).

In series circuits, you add the complex impedances of all the series elements. In parallel circuits, you add the complex admittances of all the parallel elements.

Admittances are defined as follows: $Y = \frac{1}{Z}$

$G = \frac{1}{R}$ (conductance)

$B = \frac{1}{X}$ (susceptance)

$B_C = \frac{1}{X_C} = j\omega C$ (capacitive susceptance)

$B_L = \frac{1}{X_L} = \frac{-j}{\omega L}$ (inductive susceptance)

Let's consider the following example, using a frequency of 1 MHz:

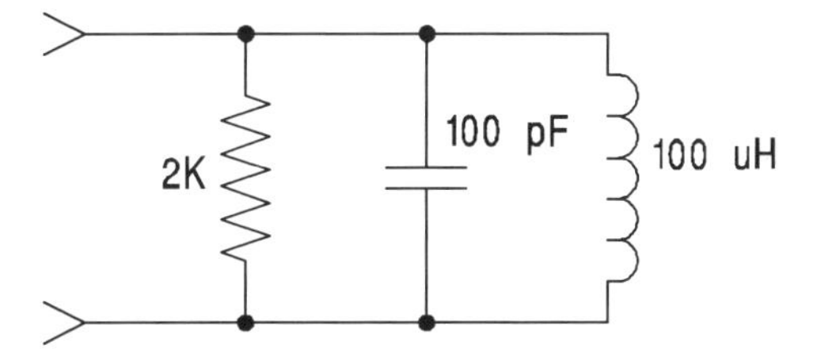

The resistor has a value of 0.5mS conductance. The capacitor has a susceptance of j6.28mS, and the inductor has a susceptance of -j1.59mS. Note that positive values represent capacitors, and negative values represent inductors: this is the opposite of reactances!

To determine the total admittance, we add the admittances of the individual parallel components, treating the real values separately from the imaginary values. Therefore :

$$Y_{total} = (5 \times 10^{-4}) + j[(6.28 \times 10^{-3}) + (-1.59 \times 10^{-3})] = 5 \times 10^{-4} + j4.69 \times 10^{-3} \text{ Siemens}$$

Another convenient way of expressing this value is to use so-called "exponential notation", where the letter "E" means "times ten to the power of". The value is therefore Y_{Total} = 5E-4 + j 4.69E-3 Siemens. Since the susceptance is positive, we know that this circuit is capacitive. Now, if we really wanted to know what the impedance of the circuit was, we have to convert the admittance, remembering the relationship Z=1/Y. However, how do you take the reciprocal of a complex value? There are two possible ways to do this:

1) Use the previously described rules for dividing complex numbers.

The first step is to convert the admittance to its polar form. The magnitude is equal to the square root of the sum of the squares, or:

$$\text{Magnitude} = \sqrt{(5.0E-4)^2 + (4.69E-3)^2} = 4.72\text{E-3 Siemens}$$

$$\text{Angle} = \tan^{-1}\left(\frac{4.69E-3}{5E-4}\right) = \tan^{-1}(9.38) = 83.915 \text{ degrees}$$

Therefore the impedance is:

$$Z = \frac{1}{Y} = \frac{1}{(4.72E-3) \text{ at } 83.915\text{deg.}} = 211.9 \text{ ohms } \angle\text{-}83.915°$$

Converting back to rectangular form:

$$\text{Real part} = 211.9 \cos(-83.915) = 22.47 \text{ ohms}$$
$$\text{Imaginary part} = 211.9 \sin(-83.915) = -210.8 \text{ ohms}$$

Therefore the impedance is **22.47 - j210.78 ohms**.

2) Use the **special formulas** for converting between admittance and impedance:

$$R = \frac{G}{G^2 + B^2} = \frac{5E-4}{(5E-4)^2 + (4.69E-3)^2} = 22.47 \text{ ohms}$$

$$X = \frac{-B}{G^2 + B^2} = \frac{-4.69E-3}{(5E-4)^2 + (4.69E-3)^2} = -210.8 \text{ ohms}$$

Therefore the impedance is again equal to **22.47 - j210.8 ohms**.

As you can see, either of the two approaches gives the same result. The particular method to use is purely a matter of personal preference.

We will now calculate the impedance of this same parallel circuit at a frequency of 503KHz. The admittance of the capacitor is j3.16 mS, and the admittance of the inductor is -j3.16 mS. The resistor's admittance is still 0.5 mS. Adding these all up, we find that the capacitive susceptances cancel out, and we are left with a total admittance for the circuit of 0.5mS +j0. Since there is no imaginary part to this value, we can invert it directly to find that the equivalent impedance is a pure resistance of 2K ohms. This is **parallel resonance**! We have just proven that at resonance, the capacitive and inductive susceptances of a parallel tuned circuit cancel out, and all we are left with is any remaining parallel resistance.

Equivalent Series And Parallel Circuits

Sometimes it is desirable when analyzing circuits to transform from parallel to series configurations. This is especially useful when matching the input impedance of VHF transistors. Examine the two following circuits:

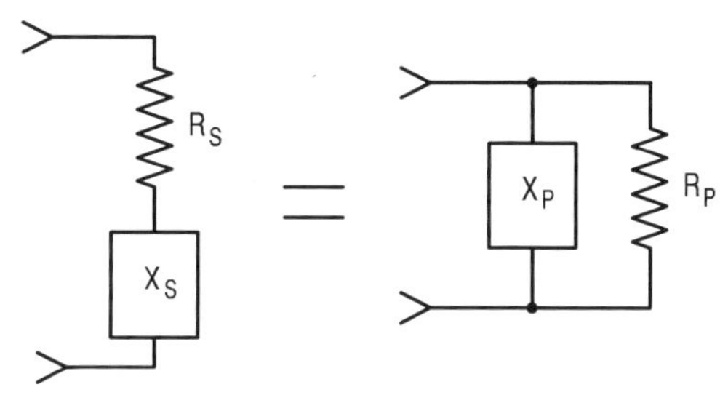

If the values are carefully calculated, it should be possible to make the two circuits act identically as far as external stimuli are concerned. There are a handy set of formulas that allow us to do this:

For converting from series to parallel -

$$R_P = \frac{R_S^2 + X_S^2}{R_S}$$

$$X_P = \frac{R_S^2 + X_S^2}{X_S}$$

For converting from parallel to series -

$$R_S = \frac{R_P\, X_P^2}{R_P^2 + X_P^2}$$

$$X_S = \frac{R_P^2\, X_P}{R_P^2 + X_P^2}$$

Let's work out an example. Assume that we have a series circuit consisting of a 10 ohm resistor and a 1000 pF capacitor. The frequency is 1 MHz. The reactance of the capacitor at this frequency is 159 ohms. We would like to know what the equivalent parallel circuit is at this frequency.

Using the above formulas, we calculate that R_P = 2.54 K ohms, and X_P = 160 ohms. We can therefore state that at 1 MHz, a series circuit consisting of a 10 ohm resistor in series with a 1000 pF capacitor is equivalent to a 994 pF capacitor in parallel with a 2.54 K ohm resistor. This is shown below:

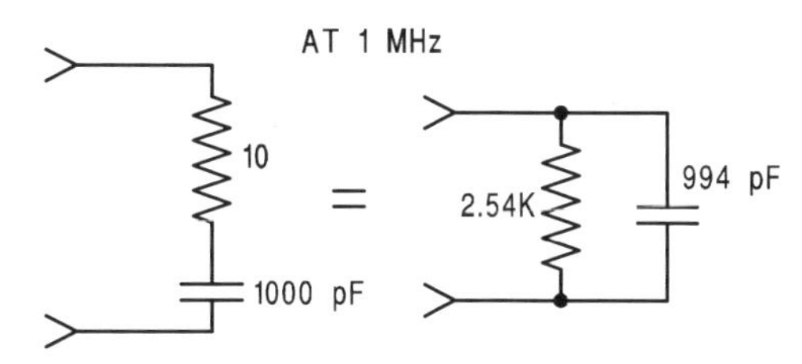

From the standpoint of their behaviour to 1 MHz signals, these two circuits are indistinguishable!

RESONANCE REVISITED

We all remember that in either a series or a parallel circuit, resonance occurs at the frequency when the capacitive and inductive reactances are equal. We will now examine the phenomena more carefully by including the effect of series and parallel resistance. The book "RF Circuit Design" by Chris Bowick gives an excellent and comprehensive review of complex impedances and resonance.

Any capacitor has some amount of leakage. This is analogous to saying that there is resistance (usually very high) connected in parallel with it. This "leakage" is not just due to the finite DC resistance of the dielectric, but is also due to the RF losses through the dielectric. The leads of a capacitor have inductance: this is usually not a problem for frequencies below 30 MHz (assuming that short leads are used), but is a major consideration at UHF and above.

An inductor has series resistance. This is a combination of the actual ohmic resistance due to the material, and the effective "RF resistance", which is the resistance to RF currents caused by **skin effect**. Recall that at radio frequencies, current flows on the outside of conductors, not through the bulk. The actual depth of the conductor that is "penetrated" by the RF current is called the "skin depth", and this is a function of frequency. At a frequency of 1MHz, the skin depth in copper is 0.007cm (2.75 thousandths of an inch). The skin depth decreases as the square root of the frequency. The effective resistance of an inductor is usually much more significant than the capacitor's effective parallel resistance in most tuned circuits.

A parallel tuned circuit can be drawn as follows, where R_L is the effective series resistance of the inductor, and R_C is the effective parallel resistance of the capacitor due to dielectric losses.

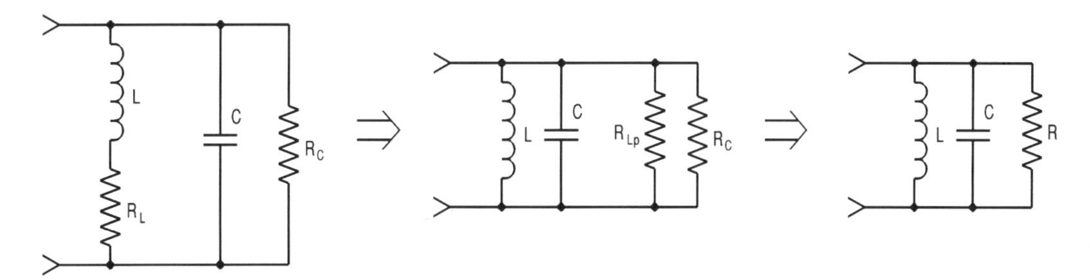

In order to analyze this circuit, we must first of all convert the series resistance of the inductor to an effective parallel resistance. This can be done using the convenient formulas given a few pages ago. We now have an inductor and capacitor in parallel with two resistors. We can easily replace the two resistors with a single resistor whose value is equal to the parallel resistance of the other two, leaving us just a single resistor, capacitor, and inductor in parallel. In order to determine the admittance of the entire circuit, we simply add the admittances of each of the three components:

$$Y_{Total} = \frac{1}{R} + j\omega C + \frac{-j}{\omega L} \quad \text{Siemens}$$

In order to determine the impedance, we simply take the reciprocal of the admittance:

$$Z = \frac{1}{Y} = \frac{1}{\frac{1}{R} + j\omega C + \frac{-j}{\omega L}} \quad \text{Ohms}$$

Now, this looks **rather messy** to handle! Let's try to simplify this expression of impedance by multiplying both top and bottom by the term "jωLR", remembering that j times j gives an answer of minus one:

$$Z = \frac{j\omega LR}{j\omega L - \omega^2 LCR + R} \quad \text{ohms}$$

This only looks a little bit easier to handle. However, it is in a very handy form. Remember that at resonance, $X_C = X_L$, and the resonant frequency is equal to the reciprocal of $2\pi\sqrt{LC}$. This is another way of saying that at resonance, 1/LC is equal to ω^2. Now, let's assume that we are at resonance, and substitute 1/LC for the term ω^2:

$$Z = \frac{j\omega LR}{j\omega L - \left(\frac{1}{LC} \cdot LCR\right) + R} = \frac{j\omega LR}{j\omega L - R + R} = R \quad \text{ohms.}$$

Wow! When the frequency is set to a value of $1/2\pi\sqrt{LC}$ Hz, the complicated looking expression simplifies to give a value of R ohms: the parallel **circuit looks just like a resistor**. If the resistance was infinite, then the impedance of this circuit would appear to be infinite at resonance: with real components however, there is always equivalent parallel resistance.

If a current source were connected to this parallel tuned circuit, and the parallel resistance was very high, then the voltage developed across the circuit would be extremely large at resonance. However, if it is desired to transfer power out of the circuit to drive other stages, a load resistance would effectively be placed across the circuit, and the voltage would not be as high.

In most tuned circuits, the losses through the inductor are more significant than those through the capacitor. A measure of the quality of an inductor is its so-called "Q". Q is the ratio of the inductive reactance to either its series equivalent parallel resistance:

$$Q = \frac{X_{series}}{R_{series}} \quad \text{or} \quad Q = \frac{R_{parallel}}{X_{parallel}}$$

The relationship between a component's series resistance and its effective parallel resistance is a function of the component's Q:

$$R_{parallel} = \left(Q^2 + 1\right) R_{series}$$

In practical circuits, a parallel tuned circuit is usually connected to some signal source having a specific source resistance R_S, and a load having a load resistance R_L:

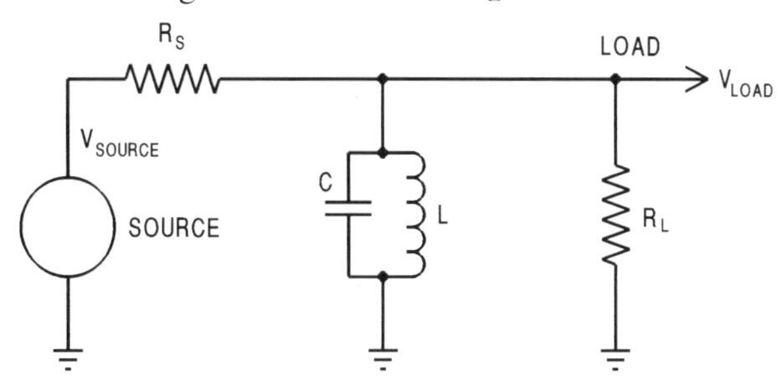

This overall circuit behaves as a voltage divider. If the parallel capacitor and inductor were removed, the overall transmission of this circuit would be independent of frequency, and would be equal to $R_L/(R_S + R_L)$. Assume for the moment that very high Q components are used, such that the effective parallel resistance of the inductor and capacitor are much higher than the source or load resistance, allowing us to ignore them for the moment. At resonance, the impedance of the parallel circuit formed by the capacitor and the inductor approaches infinity (in other words, it is an open circuit), and the voltage transfer function of the circuit is just $R_L/(R_S + R_L)$. Assume for the moment that the load resistance R_L is very high; now we can say that the signal is virtually unattenuated at resonance.

As the frequency varies either side of resonance, the magnitude of the parallel impedance of the capacitor and inductor need to be considered, and the voltage transfer function falls off. We will now plot the voltage transfer function of this circuit as a function of frequency for two different values of source resistance: 50 ohms and 500 ohms.

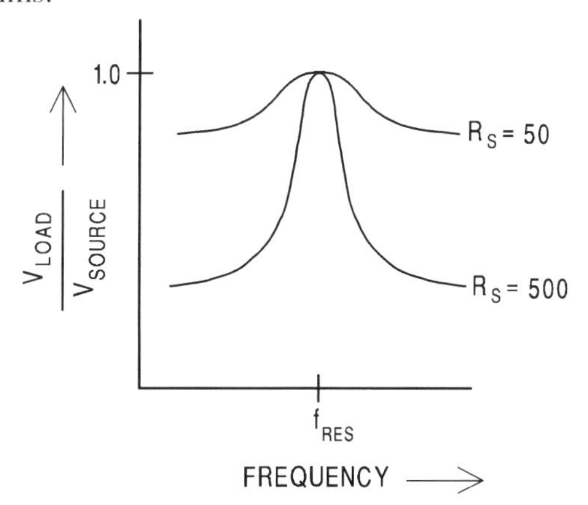

Notice that the selectivity of this circuit is much better with a high source resistance. This is because the 50 ohm source resistor caused significant loading, and serves to "de-Q" the resonant circuit. The Q of a resonant circuit is defined as:

$$Q_{circuit} = \frac{f_{centre}}{f_{3dB\ upper} - f_{3dB\ lower}} = \frac{f_{centre}}{\text{3dB Bandwidth}}$$

Note that this is the <u>circuit Q</u>, not necessarily the Q of any given component. The circuit Q can be calculated fairly simply:

$$Q_{circuit} = \frac{R_p}{X_p},$$

where R_p = the equivalent parallel resistance of R_S, R_L, and the parallel resistances of the inductor and the capacitor.

Examining this equation, it can be seen that the way to increase selectivity is to increase R_P. This implies that high Q components be used, and that the source and load resistances be high.

It is not always practical to use high source and load resistances. Indeed, these are often set by the type of transistors being used, and can be quite low. In order to get around this problem, it is possible to use impedance transforming circuits. These fool the resonant circuit into "thinking" that the source resistance is much higher than is actually present. The two most common ways of doing this are to tap down on the inductor, or split the capacitor into two series components, as shown below:

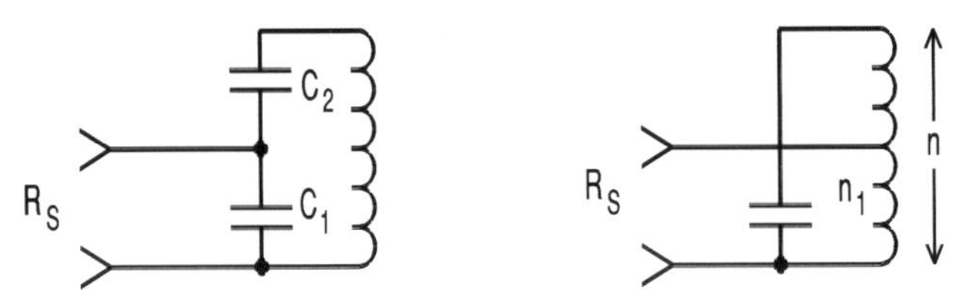

These circuits transform the actual source (or load) resistance according to the following relationships:

$$R'_S = R_S\left(1 + \frac{C_1}{C_2}\right)^2 \quad \text{For the tapped-C transformer}$$

$$R'_S = R_S\left(\frac{n}{n_1}\right)^2 \quad \text{For the tapped-L transformer}$$

In these two expressions, R'_S is the effective load resistance that appears to be in parallel with the inductor. Note that in both of these cases, the apparent resistance seen by the inductor is higher than the actual source resistance R_S.

Coupled Tuned Circuits

If component Q's are high and loading due to source and load are low, a single resonant circuit consisting of parallel components can provide relatively good selectivity. For frequencies up to about 30 MHz, it is reasonable to expect to get a circuit Q of up to 50 if the circuit is lightly loaded. However, there are many situations where a single tuned circuit is insufficient to meet particular requirements for a bandpass filter. If sharper roll off at the band edges is required (higher "shape factor"), or if better control over the passband characteristics is needed, a combination of several parallel tuned circuits can be used.

If the tuned circuits are all tuned to the same frequency, are physically separated, and are isolated by buffer stages so that there is no interaction, the overall Q is increased as follows:

$$Q_{total} = \frac{Q}{\sqrt{\sqrt[N]{2} - 1}}$$

where Q_{total} = the total Q of the overall circuit
Q = the Q of each individual stage
N = the number of tuned circuits.

Using multiple isolated tuned circuits in this manner, it is possible to get effective Q's of several hundred. If the object is to get very sharp response drop off, but a broad pass band frequency range, the resonant frequencies of each of the individual stages may be offset slightly - this is called "stagger tuning", and was a common procedure in old FM broadcast and TV receivers. More modern designs use ceramic, crystal, or SAW filters.

Rather than avoiding interaction between tuned circuits, it is possible to intentionally provide coupling means between multiple tuned circuits in order to effect a desired response. There are three primary means of providing coupling: capacitive, inductive, or magnetic. Examples of these are shown below:

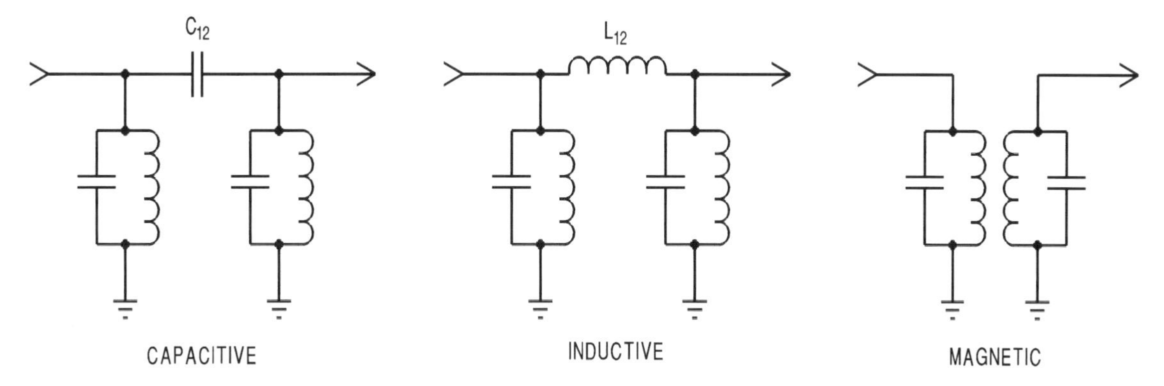

As the coupling between the circuits is increased, the overall insertion loss at resonance is decreased, until a point called "critical coupling" is reached. beyond this point, the bandwidth of the pass band is increased, while the shape of the circuit's skirts is preserved. This is shown below:

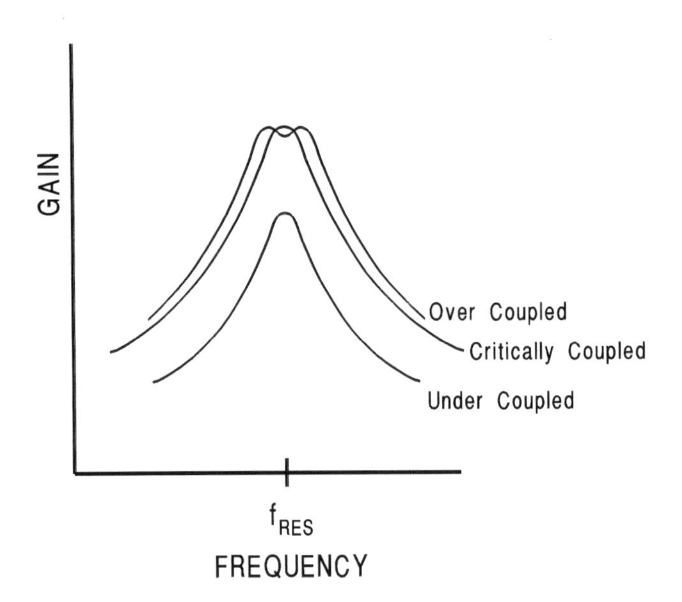

It can be seen that "over coupling" provides a similar response to "stagger tuning" in some cases. In order to provide "critical coupling" (maximum gain, consistent with no broadening of the response) between two parallel tuned circuits, the coupling capacitance or inductance should be set as follows:

$$C_{12} = \frac{C}{Q}$$

where C_{12} = the coupling capacitance
C = the resonant circuit capacitance
Q = the loaded Q of a single tuned circuit

$$L_{12} = QL$$

where L_{12} = the coupling inductance
L = the resonant circuit inductance
Q = the loaded Q of a single tuned circuit

It is interesting to note that the reactance of the capacitor and the inductor from the above equations is the same at resonance.

Magnetic coupling (often called "transformer coupling") is much harder to measure or calculate; the amount of coupling is usually determined experimentally.

FILTERS

Filters are commonly used in RF design. The main purpose of a filter is to pass a certain range of frequencies, while attenuating other ranges of frequencies. A special type of filter is also available that has minimal effect on the amplitude of the signals that are passed through it, but shifts the phase of signals in a pre-determined fashion.

Filters have traditionally been fabricated by combinations of reactive components (capacitors and inductors), but it is also possible to create a filter using purely digital techniques. Filters which are required to pass appreciable power almost always consist of inductors and capacitors in some combination. Note that at very high frequencies, individual discrete capacitors and inductors may not be observable; instead, specific lengths of transmission lines or cavities are used.

There are four basic filter types: lowpass, highpass, bandpass, and bandstop.

A filter has an input and an output. A lowpass filter allows signals with frequencies of less than a certain value to pass through relatively un-attenuated. The range of frequencies that pass relatively unmolested are called the "pass band", while those which are to be blocked are referred to as the "stop band". An ideal low pass filter would pass the desired frequencies through with zero attenuation, but would offer infinite attenuation for frequencies just higher than the top of the pass band. If plotted on a graph, the response of this filter would resemble a "brick wall": this term is sometimes used to describe high performance filter designs. The frequency response of "real" filters is less abrupt. An example of several real filter responses is shown below:

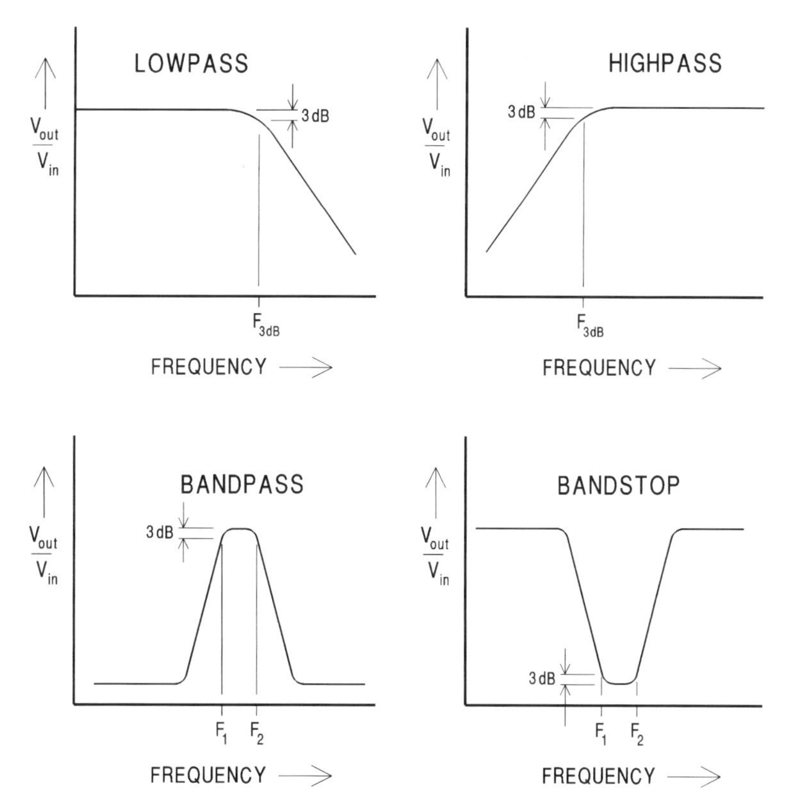

The "break frequency" is defined as the frequency at which the output has fallen to 3 dB below the level obtained in the pass-band. This is usually referred to as f_{3dB}. Any variation in the filter's attenuation in the pass-band is referred to as ripple, and is specified by the peak-to-peak variation in dB. A band-pass or band-stop filter will have two f_{3dB}s: an upper and a lower. A band-pass filter can be specified in terms of its shape factor, which is the ratio of the 60 dB bandwidth to the 3dB bandwidth, as defined in the following figure:

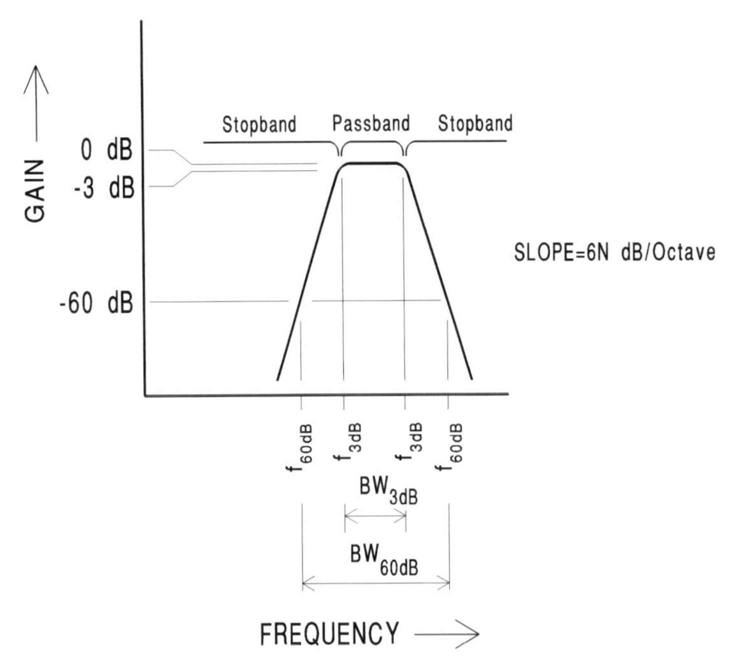

The slope of the attenuation vs frequency curve for frequencies at the start of the stop band is often an important characteristic. The slope is related to the shape factor of a band-pass filter. The slope is a function of the order of the filter, which is equal to the number of reactive components in the filter. Each reactive element contributes 6 dB per octave of slope. It is convenient to plot the response of filters on graphs that have a logarithmic frequency axis, and a vertical axis that is linear in dB (remember that dB is itself a logarithmic ratio). Some of these terms are illustrated in the following two example plots:

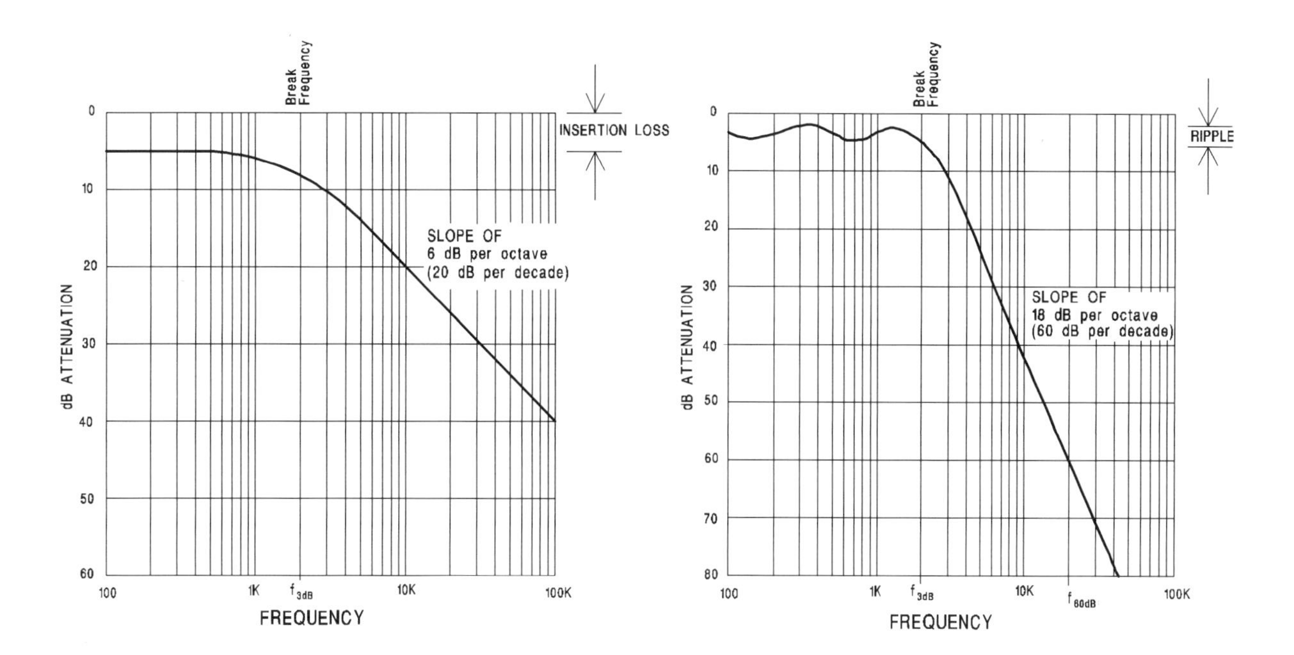

Filter design can be highly complex! Many thick books have been written about filter synthesis and analysis. We will just touch on a few of the concepts in these notes.

R-C Passive Filters

The simplest type of filter is a passive RC filter. This means that the filter consists of nothing but resistors and capacitors: there are no inductors, amplifiers, buffers, or sources of power. A simple RC low-pass filter is shown below:

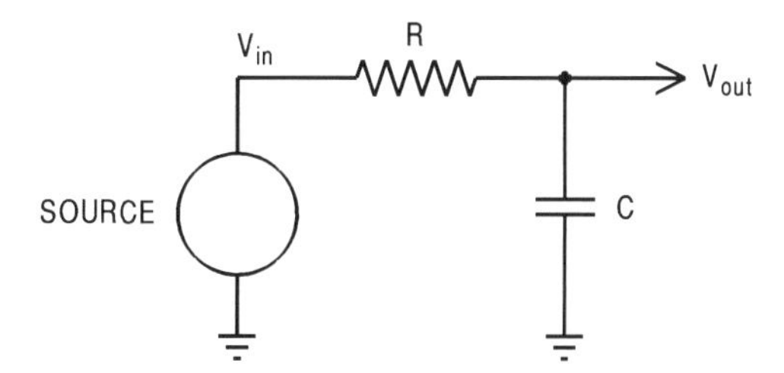

Assume for the moment that the source has zero impedance, and the load has infinite impedance. We can use the techniques previously described (remembering that j times j equals -1) to determine that the voltage **transfer function** of this circuit is simply:

$$\frac{V_{out}}{V_{in}} = \frac{\frac{-j}{\omega C}}{R + \frac{-j}{\omega C}} = \frac{1}{1 + j\omega CR}$$

Let's look at the magnitude of this transfer function separately from the angle. Note that the magnitude of the transfer function is actually defined as the circuit's gain:

$$\text{Gain} = \frac{1}{\sqrt{1+(\omega RC)^2}} \qquad \text{Angle} = -\tan^{-1}\left(\frac{\omega RC}{1}\right)$$

This is a simple circuit to analyze. We will look at the response at three different frequencies: zero, approaching infinity, and a frequency of $1/(2\pi RC)$. At zero, the term ωRC is zero, and the response of the circuit is such that there is no attenuation, and no phase shift. At a frequency approaching infinity, the gain approaches zero, and the phase shift approaches -90 degrees. At a frequency of $1/(2\pi RC)$ Hz, the gain is equal to $1/\sqrt{2}$, which is 0.707. The phase shift is equal to -45 degrees at this frequency. Converting the gain to decibels, we find that this circuit has a gain of -3dB at the "corner frequency" of $1/(2\pi RC)$ Hz. For frequencies that are higher than the corner frequency, the gain rolls off at about 6 dB/octave. (*In other words, the voltage attenuation doubles each time the frequency is doubled*). Note that a slope of 6dB/octave is equivalent to a slope of 20 dB per decade. This is a very useful characteristic to know: the response of this simple RC low pass filter can now be plotted on a **Bode Plot**, which simply shows a circuit's gain (in dB) and phase shift (in degrees) as a function of frequency.

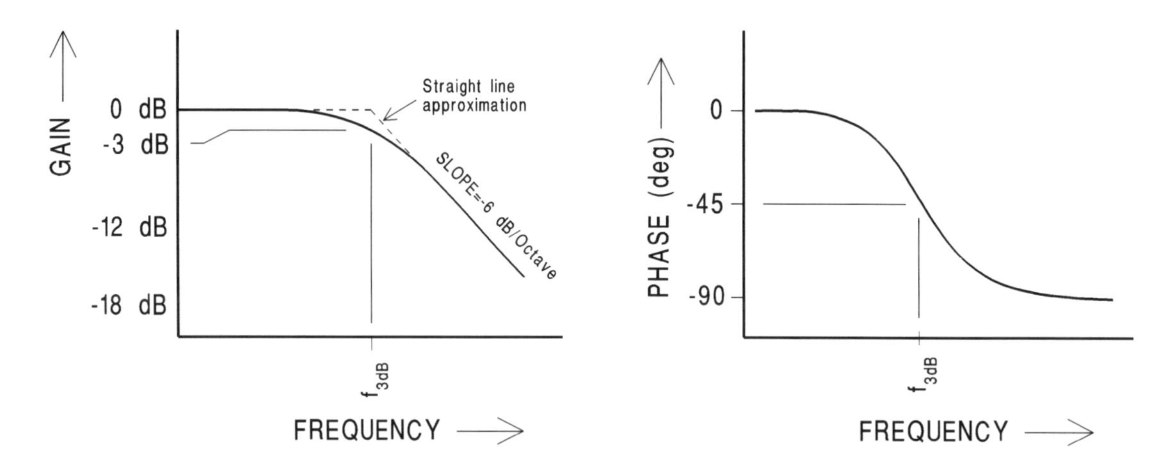

Note the straight line approximation that is used for the gain response. We will make much use of straight line approximations in other Bode Plots. If the resistor and capacitor are reversed, we can create a simple RC high pass filter. The gain rolls off at 6 dB/octave for frequencies below the corner frequency.

These two RC circuits seem very simple but unfortunately the performance is not very aggressive to be considered as a high performance filter. A roll-off of 6 dB per octave is much too modest for many needs.

In order to double the rate at which attenuation varies with frequency, we should be able to cascade 2 or more of these single RC circuits. However, we previously assumed that the load had an infinite impedance: if we connect one filter to the output of another, this assumption is no longer true. To avoid this problem we could put buffer amplifier stages between the RC stages. An easier solution would be to use larger values of R and smaller values of C (but keeping the RC product constant) on successive stages. A cascaded low-pass filter such as this is sometimes called a "Synchronously Tuned Filter". If a very large number of stages is used, the filter becomes a Gaussian Filter. An example of a three stage Synchronously Tuned Filter, together with its magnitude response is shown below:

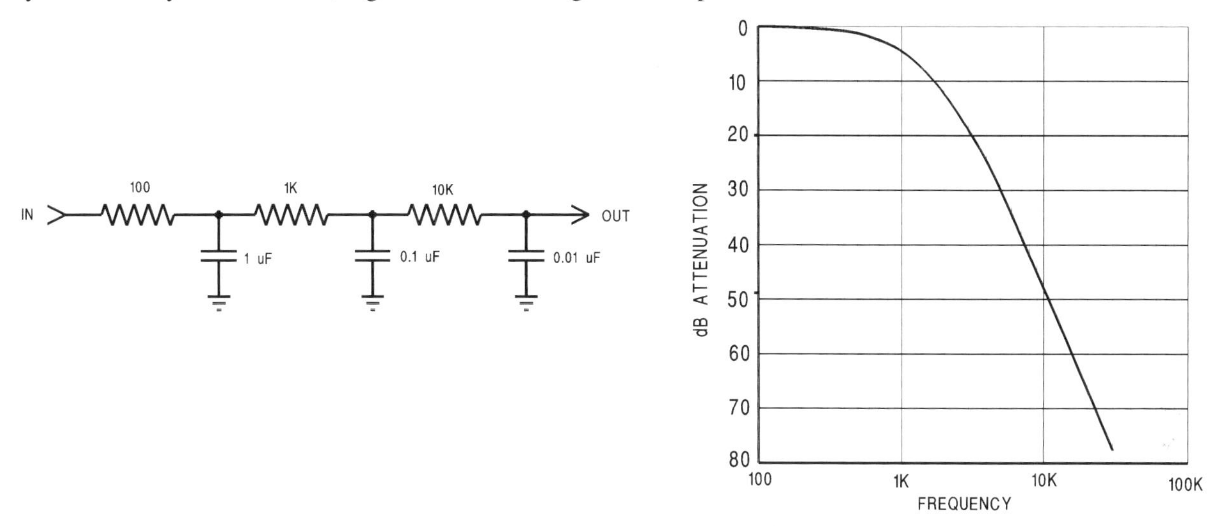

Although this makes a very simple structure, the resulting performance still leaves something to be desired. At the break frequency, we now have 3 dB loss from <u>each</u> of the RC stages, and the roll off is much too gradual for most applications. In looking at the values used, the expected 3dB break frequency for an individual stage is about 1.6 KHz, but when the three stages are cascaded, the effective 3dB break point moves down to under 900 Hz in this example.

RC Active Filters

It would be nice to have a filter circuit that very quickly makes the transition from the relatively flat passband to the ultimate roll off slope (in other words, we would like to have a good shape factor). Using combinations of inductors and capacitors, filters with very good shape factors can be designed. If inductors are not used in a filter design, the only way to get excellent filter shapes is to add an active component (something that adds energy back in to the system). Operational amplifiers are often used in the design of low frequency filters.. As soon as active components (such as op amps) are used as an integral part of a filter's structure, the circuit is referred to as an "**active filter**".

We will try to avoid introducing the concepts of **complex frequencies, the S-Plane,** and **pole-zero diagrams** at this point - it is hard enough to deal with just complex impedances! Because of this simplifying assumption, the following section may be a little bit "mechanical". Those of who you are more mathematically inclined might like to know that the following circuits place the filter's poles in prescribed geometric arrangements in the complex frequency plane (the "s-plane").

Sallen and Key developed a simple circuit which allows the 12 dB/octave roll off of two cascaded RC stages, but keeps a reasonably sharp 3 dB knee. We will refer to this type of filter design as the SK filter.

This circuit uses an op amp as a unity gain buffer with extremely high input impedance and very low output impedance. A feedback path is used to "sharpen up" the corner point. Shown below are the two classical SK designs for a 2 pole filter.

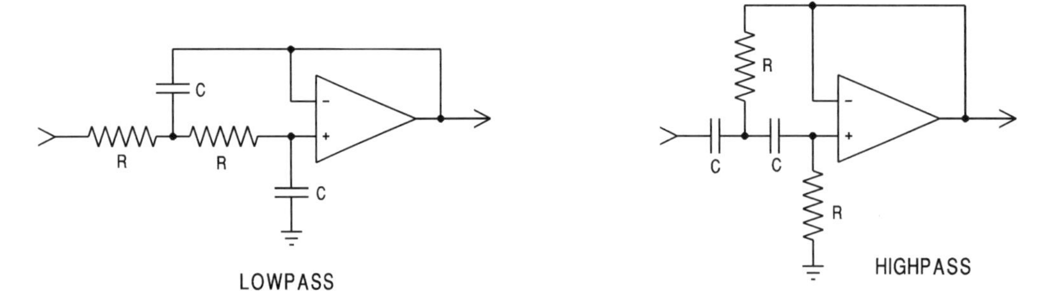

Because there are two reactive components (capacitors in this case), the ultimate attenuation rate outside of the passband is 12 dB per octave. Multiples of the above stages can be cascaded together to increase the roll off rate, but the 3 dB knee again gets "soft" unless the values are moved around in a specified fashion *(the damping ratios of the individual stages are changed, or the poles are moved out in the s-plane).*

A variant of the SK low pass filter offers three poles in a single stage, generating a roll off rate of 18 dB per octave. An example of the circuit (tuned for 1.6KHz) is shown below, together with the actual measured gain response for three different values of the feedback capacitor:

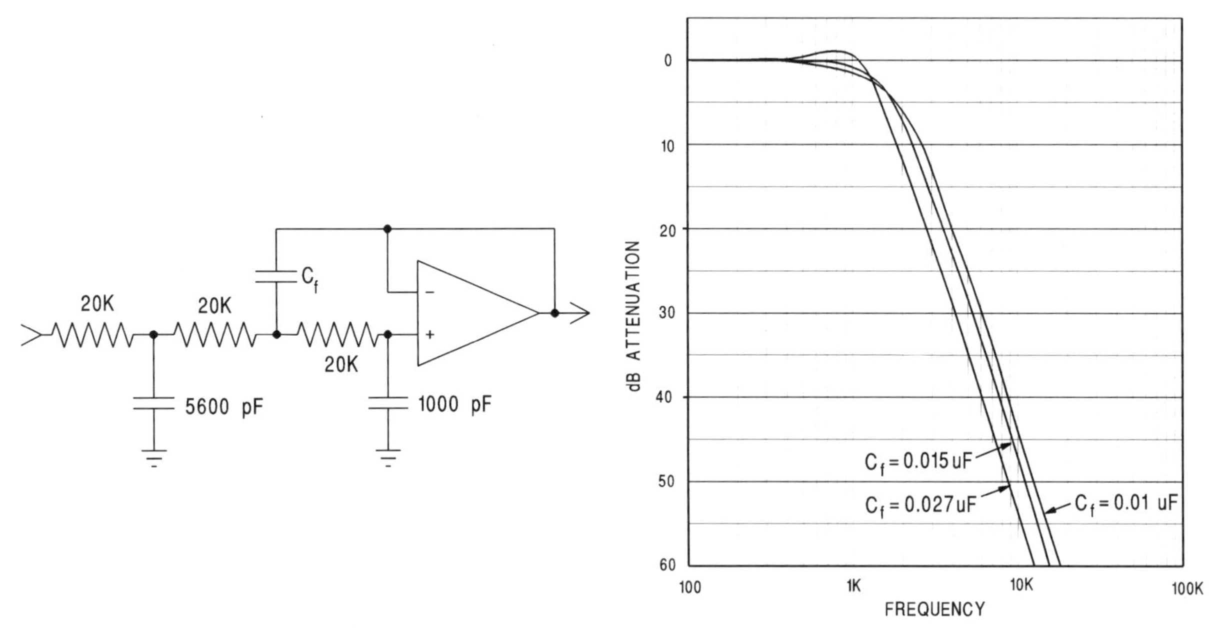

Note that by varying the feedback capacitor C_f, different responses can be obtained. With larger values, the response peaks slightly before rolling off rapidly. The peaked response gives a better shape factor, but the filter will then exhibit a few percent overshoot if a square wave is fed through it. This is a trade-off that is often made in order to get a sharp roll off and steep skirts, and is symptomatic of the Chebyshev response. Note that all of these responses have a much better shape factor than the synchronously tuned three stage RC filter described previously.

The SK filter uses the op amp as a unity gain, high impedance buffer. By slightly modifying the SK circuit to put a voltage divider in the negative feedback path, we create what is known as a **VCVS** filter (voltage-controlled voltage source filter). The advantage is that all of the capacitor values are the same. The gain of each stage is slightly greater than 1. Circuits for low-pass, band-pass, and high-pass filters are shown below:

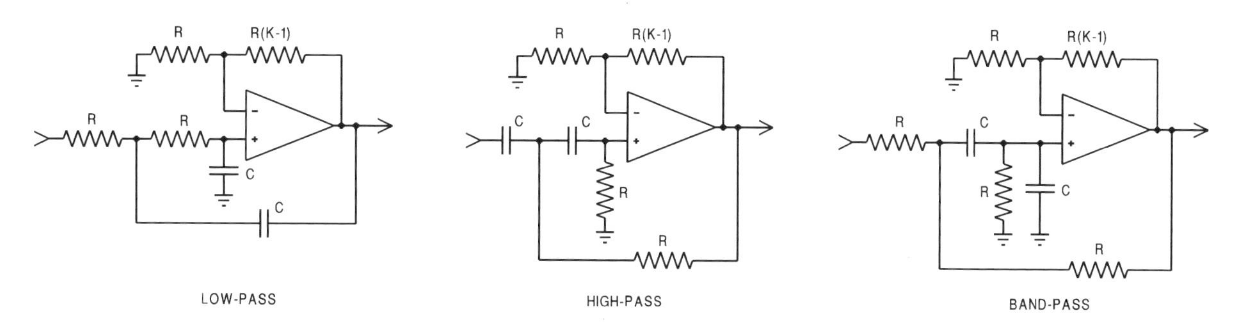

Note that the op amps could have been configured for unity gain rather than the arrangements shown above, but then the resistance (or reactance) of the feedback element would have to be increased by the factor K from the table below in order to duplicate the desired response.

Each of the above stages is a "second order" design (mathematically, the transfer function has two poles), which means that it offers 12 dB/octave roll off per stage. Multiple stages can be cascaded without compromising the sharpness of the break point by using appropriate values. The following table presents data that can be used to cascade 1, 2, 3, or 4 VCVS stages:

POLES	BUTTERWORTH	CHEBYSHEV (0.5 dB)		CHEBYSHEV (2.0 dB)	
	K	f_n	K	f_n	K
2	1.586	1.231	1.842	0.907	2.114
4	1.152	0.597	1.582	0.471	1.924
	2.235	1.031	2.660	0.964	2.782
6	1.068	0.396	1.537	0.316	1.891
	1.586	0.768	2.448	0.730	2.648
	2.483	1.011	2.846	0.983	2.904
8	1.038	0.297	1.522	0.238	1.879
	1.337	0.599	2.379	0.572	2.605
	1.889	0.861	2.711	0.842	2.821
	2.610	1.006	2.913	0.990	2.946

The names "**Butterworth**" and "**Chebyshev**" refer to the mathematical response of the filter's transfer function. Butterworth filters offer very flat frequency response in the passband, while Chebyshev filters compromise passband flatness for much sharper roll-off in the stop band. Chebyshev filters are specified by the amount of ripple that they exhibit in the pass band.

It is easiest to describe how to use this table if a Butterworth filter is discussed. Each stage uses the same value of R and C, such that the corner frequency is equal to $1/(2\pi RC)$ Hz. The gain of each stage is set according to the value of K from the above table.

The Chebyshev design procedure is slightly more complicated. Each stage's gain is set according to the specified value of K, then the particular value of the RC constant in each stage is modified from the overall filter's corner frequency by multiplying by the value of f_n given in the table. For high pass filters, the reciprocal of f_n should be used.

A bandpass filter that has enjoyed popularity for many years is the **Wein Bridge Filter**. A modern version of this structure implemented with a single operational amplifier is shown below:

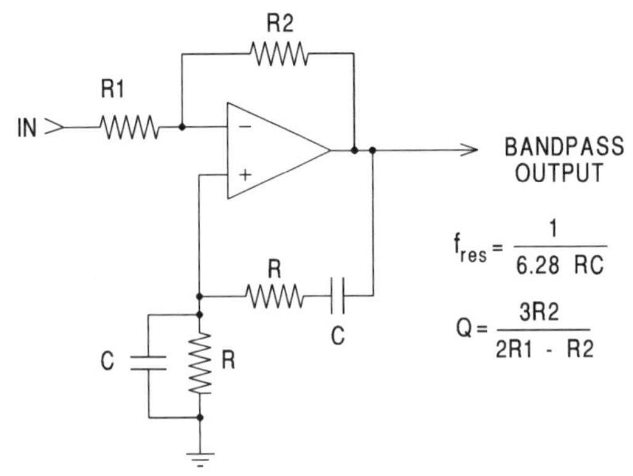

Narrow bandpass filters for low frequencies are best implemented using either the **Biquad** or **State Variable Filter** approach. This type of design uses more op amps, but it is easier to design and adjust, is more stable, and can be configured to be any of the three basic filter types. It is possible to purchase a complete State Variable Filter as an IC: you only have to add three resistors to set the gain, frequency, and Q. The basic schematic for a State Variable Filter is shown below:

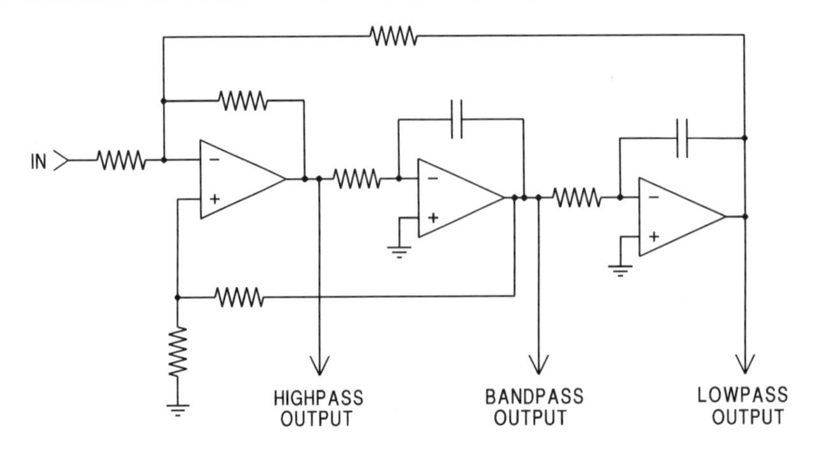

Active filters are a fascinating subject. Most of the design work has occurred since the mid-60's. A fairly easy to understand reference on this topic is contained in Chapter 5 of "The Art Of Electronics" by Horowitz and Hill.

Switched Capacitor Filters

Active filters such as those described in the previous section can give very good performance, but they require accurate component values and are generally unsuited to variable frequency operation, especially if multiple stages are involved. A low cost solution known as the switched capacitor filter (or "SCF") addresses both of these issues. An SCF is an integrated circuit device that uses internal digital FET switches to alternatively charge internal capacitors, and then dump them into successive stages. The important parameter is the ratio of the internal capacitor values, and these are easy to control in production. The switching action is controlled by a digital clock signal that is applied to the IC. The clock is usually at a multiple of the filter's break point (such as 100 times).

In order to prevent unwanted signal artifacts at the SCF's output, the signal must first be run through an external RC low pass filter to ensure that there is no energy near the switching frequency. A similar low pass filter is required at the output to remove switching noise. Since the clock signal frequency is typically 100 times higher than the frequencies of interest, a simple single stage RC filter is sufficient in both cases.

Since the SCF operates by alternately charging and discharging internal capacitors, some noise is generated, and the overall circuit's dynamic range is restricted to a little more than about 80 dB.

SCF's are available either internally configured for the desired response (low-pass, band-pass, etc.) or in a general purpose form which can be structured as required. These filters are available from several vendors, and are fairly low cost.

Passive L-C Filters

Active filters and passive RC filters are most suited for audio frequencies. At radio frequencies traditional LC filters rule the waves, especially if any amount of power has to be passed through. Either Butterworth or Chebyshev responses are the most commonly employed.

The chapter entitled "Filter Design" in the excellent reference book "RF Circuit Design" by Chris Bowick (WB4UHY) contains a wealth of information on LC filter design, including details on designing Butterworth and Chebyshev filters with unequal load and source resistances. Design tables exist for low-pass filters normalized for a 1 radian/second cutoff frequency and a 1 ohm source impedance: various scaling and transformation procedures are required to get the desired result.

Recent editions of the ARRL Handbook contain quite readable sections describing the design of Butterworth, Chebyshev, and Elliptic filters for 50 ohm systems over the 1 to 10 MHz range using standard value capacitors. Scaling procedures are described for other frequencies. We will work out an example using the tables contained in the 1994 Handbook.

For our example, we will design a low pass filter that can provide moderate attenuation of harmonics of a typical "low band rig" (1 - 30 MHz). We will use a 7th order Chebyshev filter with 0.1 dB ripple in a Pi configuration. In order to minimize any effect on signals at the high end of the amateur 10 metre band, we will set the break frequency at 32 MHz. Table 21 in the Handbook can be used to directly read off the values for the capacitors and inductors for a 1 MHz filter: we have to scale these values to our frequency requirement by dividing them all by 32. The final filter design will therefore look like:

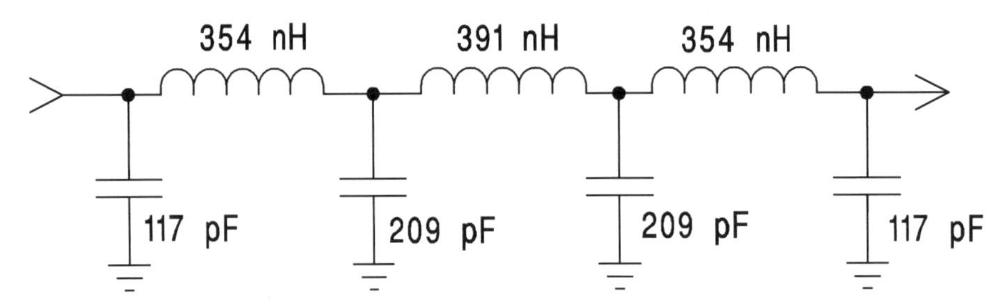

In order to be effective, this filter should be built in a shielded enclosure. Magnetic coupling between the inductors must be avoided either by using physical separation or by using toroidal cores.

Other Filter Designs And Responses

Earlier in this chapter we discussed the Butterworth and Chebyshev filter responses. The Chebyshev filter trades off passband ripple for a faster roll-off into the stopband region. There are actually many more filter responses available to those who are comfortable with the mathematics involved in their design. The figures below show characteristics of several of these as implemented in a 1 MHz low pass filter (5th order):

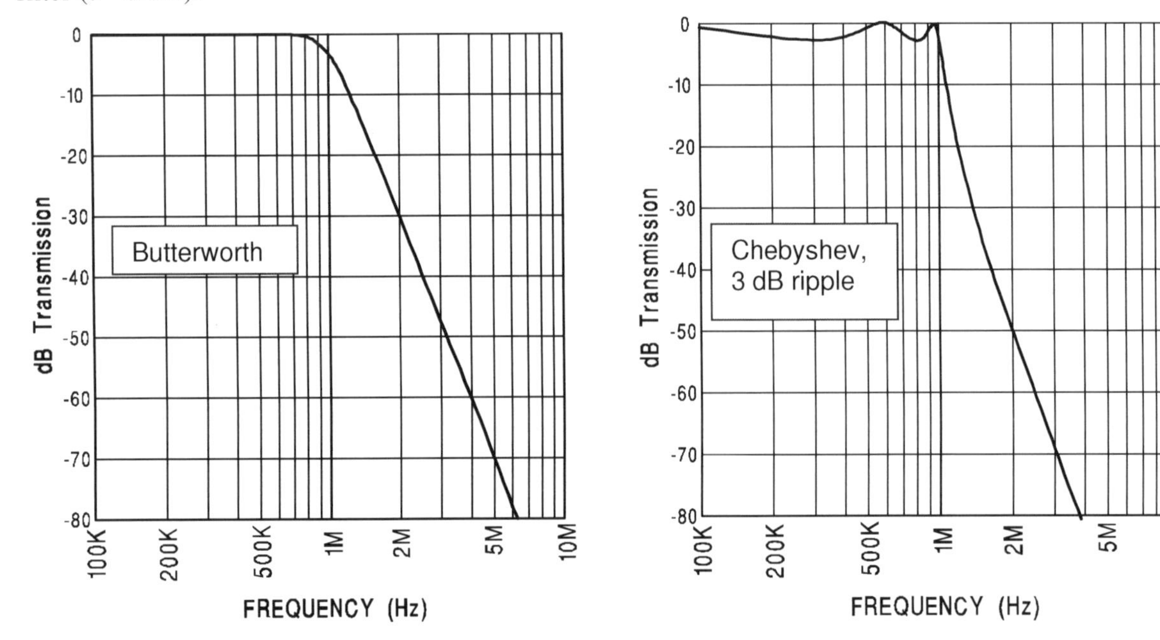

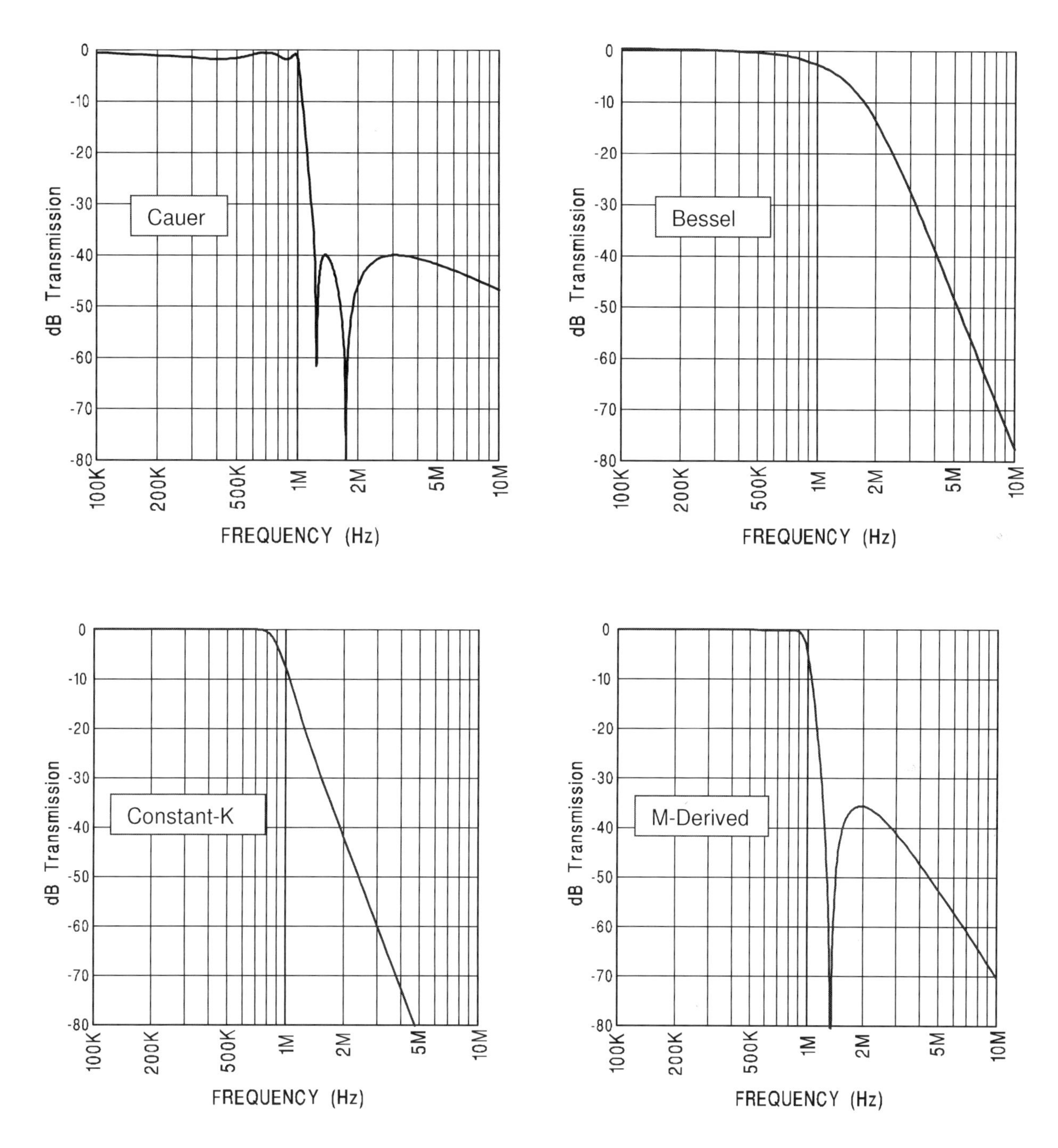

Note that for each of these filter designs, we are only showing the transmission function (the "gain") as a function of frequency. Each has its own characteristic phase response, and some are preferred for certain applications because of the way that they preserve the fidelity of transmitted pulses.

Commercial software is available to quickly and accurately design LC filters of any configuration and response. One easy-to-use program is called "**Elsie**", and the student version of this may be downloaded for free from Tonne Software.

Higher Frequencies

For frequencies over about 400 MHz, microstrip and/or strip line structures can be used to implement filters. These are described in a later chapter. For VHF and UHF frequencies, "cavity filters" allow very high Q's to be implemented. Here is a typical design:

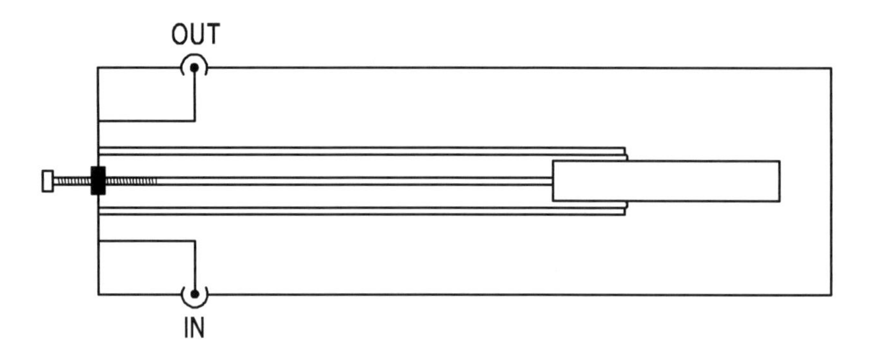

Note that this is actually a coaxial configuration, and the above drawing is intended to illustrate a cross-section taken through the centre. Cavity filters are commonly used in duplexers that allow multiple radios (both receivers and transmitters) to share a common antenna (as might be the case at a repeater).

A special form of coupled filter known as the "helical filter" is also used to achieve high performance at VHF and UHF in a relatively compact space (as compared to cavities).

Digital Filters

Digital filtering techniques only became practical in the 1980's with the advent of the DSP IC (Digital Signal Processor). The most common family of DSPs at that time was the TMS 320Cxx family from Texas Instruments. A DSP is nothing more than a dedicated processor that is designed to handle certain operations (such as multiply and accumulate) at a very high rate. As an example, the TMSC30 handled 32 bit floating point values, and could perform over 10 million multiplications per second, but more modern DSP designs from Texas Instruments, Analog Devices, and other semiconductor vendors can achieve thousands of times more performance.

A digital filter works on sampled data. Rather than feeding in a continuously varying analogue voltage, a digital filter is given a sequence of values representing the input voltage at specific points in time. The input signal is sampled at a rate that is at least twice that of the highest frequency component, and then an analogue-to-digital (A/D) converter turns the sample's value into a digital value that the filter operates on. The output of the filter is composed of values that are mathematically derived from prior values of the input. By using appropriate mathematical functions, almost any filter type and response can be created. The output values are converted back to analogue form by a digital-to-analogue (D/A) converter, and then a low pass filter is used to remove sampling noise. A block diagram of a general purpose digital filter is shown below:

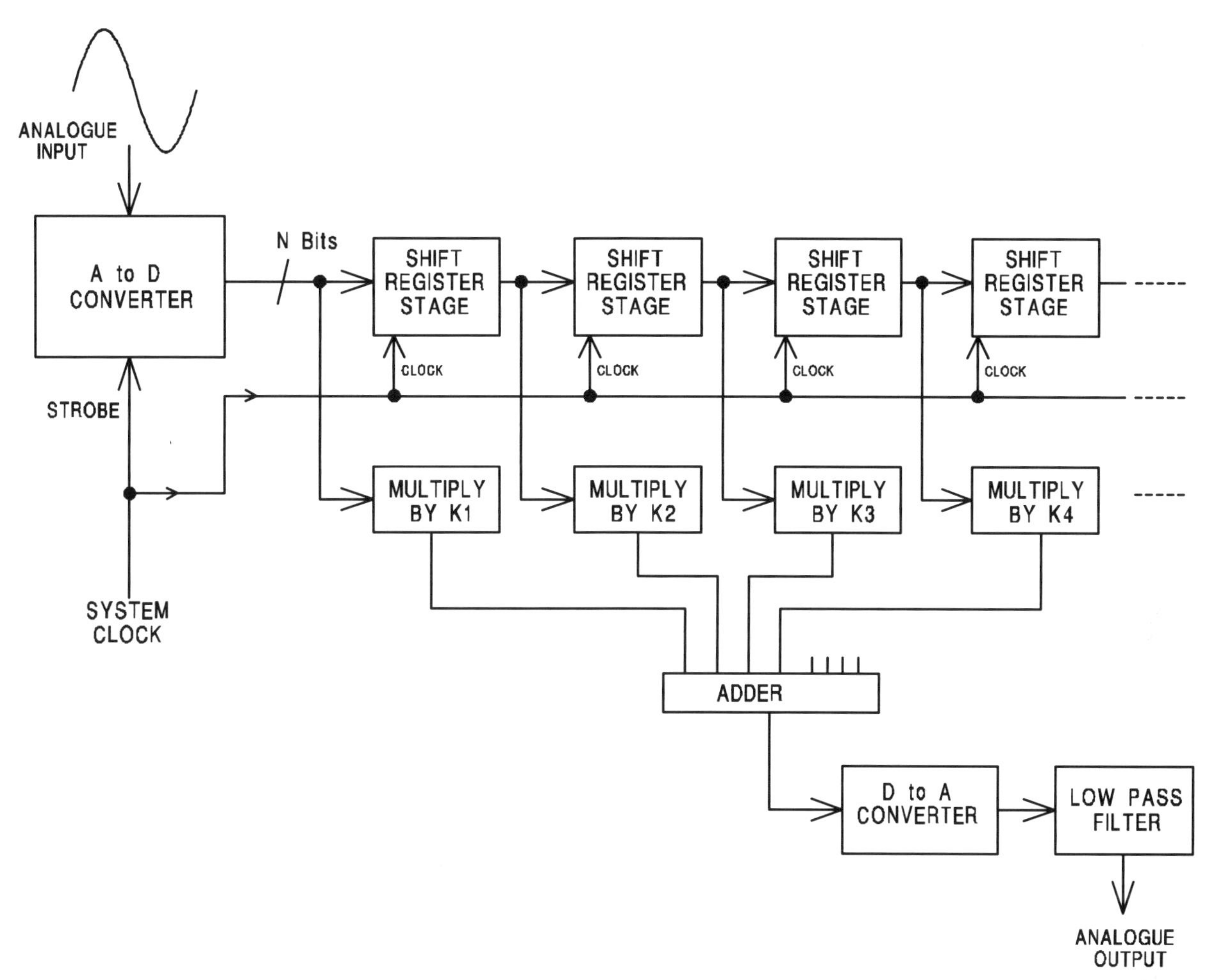

The constants K1, K2, K3, etc. determine the response of the filter. The shift register stages transfer a given value from the A/D converter to the right one step for each cycle of the system clock. So as you go further along the string of shift registers (to the right), they contain sample values of the input voltage from further back in time. All of the functions between the output of the A/D converter and the input of the D/A converter can be implemented in an appropriately programmed DSP. Input frequencies are limited to a few hundreds of KHz unless very expensive DSPs are used. It is also possible to implement all of the above blocks using discrete ICs, or by suitably designing a custom IC with appropriate circuitry.

The filter block diagram illustrates what is referred to as a Finite Impulse Response (FIR) filter. If the adder block also had inputs from delayed versions of the filter's output, the filter would be called an Infinite Impulse Response (IIR) design.

TRANSMISSION LINES

The primary purpose of a transmission line is to couple RF energy from a source to a load. All radio stations have at least one transmission line in the system: the one connected to the antenna. In an ideal situation, all of the power that enters the transmission line will be delivered to the load, and no radiation takes place along the line.

In order to prevent radiation from a transmission line, the physical arrangement of the conductors is arranged so that the electromagnetic fields cancel out. One easy-to-understand way that this can be accomplished is to use two very closely spaced parallel conductors which are passing equal signals of opposite phases. At any appreciable distance from the conductors, both the magnetic and the electric fields cancel out, and very little energy is radiated.

If a transmission line does not itself permit much radiation, the converse is true: the line will not "pick up" local signals (such as electrical noise or other transmissions) that are in the vicinity. This is a useful property to have if the transmission line is intended to connect a transceiver to a remotely-located antenna.

Transmission lines all have an associated characteristic impedance. This is sometimes a hard concept to grasp, as you don't normally think of the wires connecting a light bulb to a battery as being a transmission line with a characteristic impedance. One very unscientific way to imagine this concept is to simply think of an infinitely long transmission line: the energy never gets to the other end. Inject a signal into the infinitely long transmission line, and look at the relationship between the voltage and current: using ohm's law, the quotient will be equal to the characteristic impedance. In actuality, the characteristic impedance is defined by the inductance and capacitance of the transmission line per unit length:

$$Z_o = \sqrt{\frac{L}{C}}$$

where L and C are the line's inductance and capacitance per unit length.

The values of L and C are functions of the cross sectional physical dimensions, and characteristics of any dielectric that may be present.

In free space, radio signals travel at the speed of light, which is approximately 300,000 Km/second. In a transmission line, the speed is somewhat slower. The ratio of the actual speed in the transmission line to the speed of light in free space is called the **velocity factor**, which is a number less than one.

Often specific lengths of transmission line are used as circuit components. An important parameter is the "electrical length" of the line in terms of fractions of a wavelength. In order to determine the correct length, it is necessary to multiply the desired free space length by the velocity factor.

Transmission lines have the desirable characteristic that if a load is connected to the far end of a transmission line, and its resistance is equal to the line's characteristic impedance, then all of the power is absorbed by the load, and none is reflected back.

At radio frequencies, the current flow is not uniformly through the bulk of the conductor, but in a very shallow region on the surface of the conductor. As discussed in the section on resonance, this "skin depth" is frequency dependent: higher frequencies have reduced skin depth.

Parallel lines

The simplest transmission line consists of two parallel conductors. Assuming that equal and opposite currents are instantaneously flowing on the two conductors, then very little radiation will occur from the transmission line itself. The majority of the current will be on the "near" surfaces of the two conductors. If open line is used, there is almost no dielectric material between the conductors, other than the occasional insulator: therefore the velocity factor is close to one.

Open wire lines have very low insertion loss, even at high frequencies. However, they are not inherently shielded, and are affected by the presence of conductive or magnetic objects. Unless closely-spaced insulators are used, it is hard to maintain a perfectly constant spacing between the conductors. 300 ohm TV "twin lead" keeps the spacing constant by encasing both conductors in a single plastic strip, but the plastic causes some attenuation at high frequencies because of the dielectric's loss.

A cross section of an open wire line is shown below:

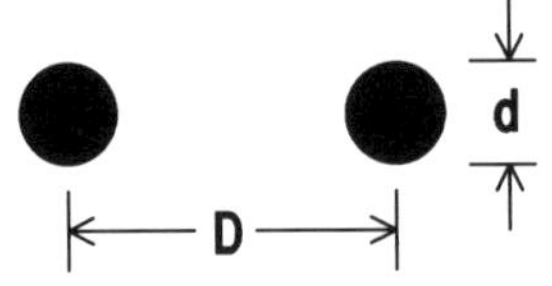

Assuming the use of open wire line, the characteristic impedance of parallel conductors can be calculated as:

$$Z_o = 276 \log\left(\frac{D}{d} + \sqrt{\left(\frac{D}{d}\right)^2 - 1}\right) \text{ ohms}$$

Here, D is the spacing between the conductors, and d is the diameter of one of the conductors.

If (as is usually the case), the spacing D is much greater than the conductor diameter d, then the impedance can be approximated as:

$$Z_o \approx 276 \log\left(\frac{2D}{d}\right) \text{ ohms}$$

If the spacing between the conductors is much less than a wavelength, little radiation will occur from open lines. However, this is a hard condition to meet at UHF and higher frequencies.

Because so little of the volume containing the electrical field is occupied by dielectric other than air, both twin lead and open air lines have relatively high velocity factors, and correspondingly small high frequency absorption loss.

Coax

Coaxial cable consists of an inner conductor mounted in the centre of a circular outer conductor. The space between the two conductors can either be mostly air, or it can be filled with a dielectric material. Lower losses at very high frequencies are possible if the dielectric is mostly air.

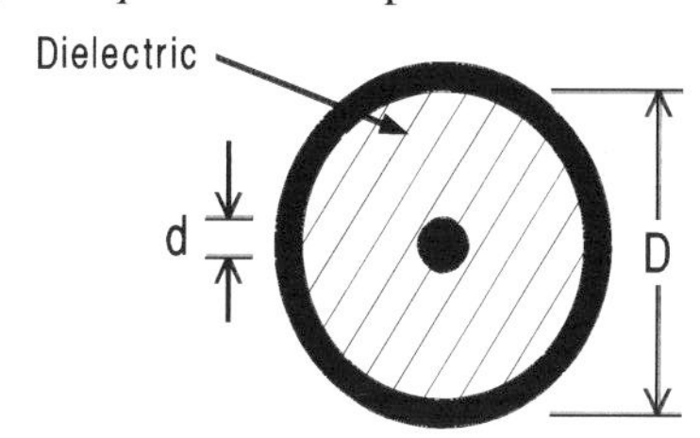

Current flowing on the inner conductor is balanced by an equal and opposite current flowing on the inside surface of the outer conductor. Because of the small skin depth at RF frequencies, no current will flow on the outer surface of the outer conductor.

If the outer conductor is solid, the coax cable is completely shielded, and no radiation will occur. Flexible coax lines are usually constructed with a braided outer conductor, and the shielding will not be perfect. To counter-act this, high quality lines often have two outer shields.

Rigid coaxial lines such as "Hardline" or "Heliax" have very little dielectric material between the two conductors. They therefore have fairly high velocity factors, and low high frequency losses.

The characteristic impedance is a function of the relative diameters of the two conductors. The only important dimensions are the outer diameter of the inner conductor, and the inner diameter of the outer conductor.

The characteristic impedance is calculated as:

$$Z_0 = \left(\frac{138}{\sqrt{\varepsilon}}\right) \log\left(\frac{D}{d}\right)$$

where:
- D = ID of outer conductor
- d = OD of inner conductor
- ε = relative dielectric constant (air = 1)

Characteristics of some common commercial coaxial cables (diameter in inches, impedance in ohms) are:

TYPE	O.D.	Z_0	ATTENUATION PER 100 FEET (DB) 1 MHz	10 MHz	100 MHz	1 GHz	3 GHz
RG-8A	0.405	50	0.16	0.55	2.0	8.0	16.5
RG-8X	0.242	50				13.5	28
RG-58	0.195	50	0.33	1.25	4.65	17.5	37.5
RG-59A	0.242	73	0.34	1.10	3.40	12.0	26.0
RG-174A	0.100	50	1.3	3.9	12		
RG-213	0.405	50	0.16	0.60	1.90	8.0	
RG-400	0.195	50				13	28
9913	0.405	50	0.13	0.41	1.30	4.3	

It should be noted that coaxial cable's shielding and non-radiating properties do not exist for **unbalanced** currents (ie those currents which only flow on the outside of the coax. To prevent these currents, balanced loads such as centre-fed dipoles should be interfaced with baluns.

Stripline

Stripline (also known as "Triplate"), can be thought of as rectangular coaxial cable with the sides removed. It is impractical to consider using stripline for conveying power to a remotely-located antenna, but it is commonly used at high frequencies (above 3 GHz) to connect signals on printed circuit boards, or to fabricate passive components such as filters, couplers, and mixers. A cross section of a stripline transmission line is:

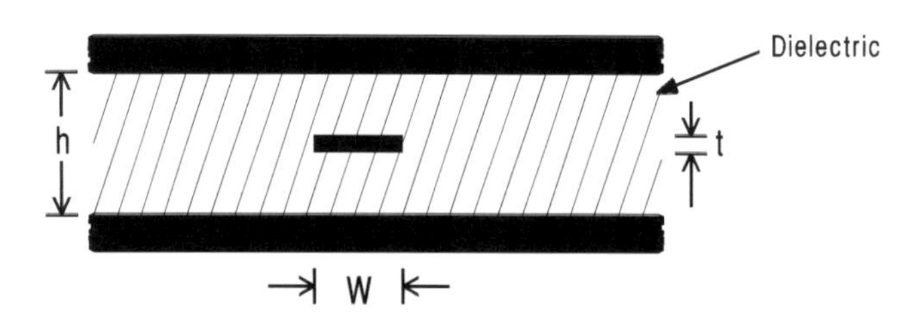

Note that stripline consists of a flat central conductor symmetrically placed between two conducting ground planes. The space surrounding the central conductor is normally filled with a dielectric material, although it could conceivably just be air. Stripline is normally made by etching printed circuit board material, then clamping them together: because of this, the dielectric is composed of the PCB's insulating material. Common printed circuit board material (such as G-10 or FR-4) uses a fiberglass-epoxy inner layer with a dielectric constant of approximately 4.8. At frequencies above 2 GHz, the losses become substantial if general purpose PCB material is used, so special board material based on Teflon-fiberglass (dielectric constant of about 2.5) or Epsilam-10 (dielectric constant of about 10) is used.

The characteristic impedance of stripline is hard to model, but the following expression gives an approximation that is within a few percent:

$$Z_0 = \left(\frac{94.2}{\sqrt{\varepsilon}}\right) \ln\left(\frac{1+\frac{W}{h}}{\frac{W}{h}+\frac{t}{h}}\right)$$

Where
W = width of the centre conductor
h = spacing between the ground planes
t = thickness of the centre conductor
ε = relative dielectric constant

The following chart can be used to determine the characteristic impedance based on the stripline's geometry:

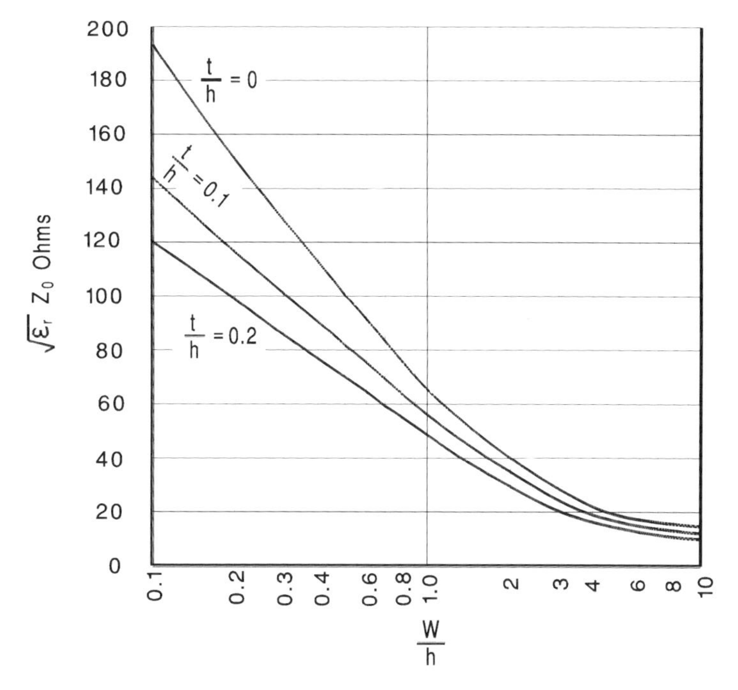

Assuming that the two ground planes extend a substantial distance either side of the centre conductor, stripline is inherently shielded. Stripline is commonly fabricated by clamping together two PCB's (one double-sided, one single-sided), however it is hard to connect components to the centre conductor because part of the dielectric from one of the PCB's will have to be machined away.

Microstrip

Because of the complexity associated with fabricating and using stripline, most amateurs prefer to work with **microstrip**, which is just like "half a stripline". Microstrip can be easily made by etching one side of a double-sided printed circuit board.

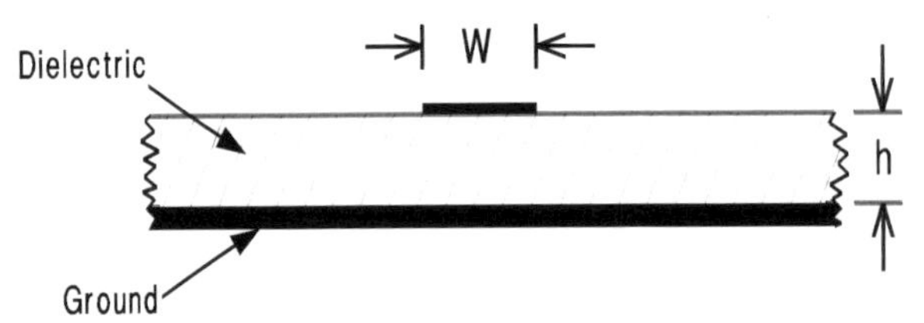

Conventional Fiberglass-epoxy board material is commonly used at frequencies below about 2 GHz, but better performance will be realized at high frequencies by using Teflon board materials.

The characteristic impedance of microstrip can be approximated as:

$$Z_0 = \frac{120\,\pi}{\left(\frac{W}{h}+1\right)\sqrt{\varepsilon+\sqrt{\varepsilon}}}$$

Where
- W = width of the centre conductor
- h = spacing between the ground plane and the centre conductor
- ε = relative dielectric constant

The following data gives typical dimensions for several different standard board types, and is accurate at 1 GHz. The values of W are in mils (0.001 inches), and the dielectric constants for Fiberglass and Teflon are 4.8 and 2.55 respectively. Z_0 is measured in ohms:

	1/16 inch Fiberglass		1/32 inch Fiberglass		1/16 inch Teflon		1/32 inch Teflon	
Z_0	W	Velocity Factor	W	Velocity Factor	W	Velocity Factor	W	Velocity Factor
20	362	0.50	172	0.50	487	0.66	231	0.66
30	221	0.51	105	0.51	305	0.67	145	0.67
40	151	0.52	72	0.52	214	0.68	102	0.68
50	109	0.52	52	0.52	159	0.68	76	0.68
60	81	0.53	39	0.53	123	0.69	58	0.69
70	61	0.54	29	0.54	97	0.70	46	0.70
80	46	0.54	22	0.54	78	0.70	37	0.70
100	25	0.55	12	0.55	50	0.71	24	0.71

Because microstrip is not completely shielded, some radiation losses do occur. If a shield is added to a microstrip component or transmission line, it must be high enough to avoid interfering with the electrical fields: one half inch is usually sufficient.

High frequency digital circuitry (greater than 50 MHz clock rates) can encounter RF-like problems as signals are passed between IC's. These "glitches" are often caused by undetected reflections or "ringing". Many high frequency digital board designs use properly terminated microstrip lines to interconnect signals: this is especially important when using ECL logic.

MICROSTRIP COMPONENTS

A variety of components are easily made using microstrip techniques. Since the microstrip lines can be fabricated virtually for "free" if a PCB is going to be used anyway for other circuitry, you might as well use the copper traces to act as transmission lines. Note that most of the following information is really only suitable for frequencies of over 500 MHz unless you don't mind physically large boards.

As we saw from the previous table, it is easy to create transmission lines with specific impedances using microstrip. Knowing the velocity factor, we can then determine how long a length of microstrip line is required to create a given electrical length in terms of fractions of a wavelength at a given frequency.

A microstrip section that is one quarter wavelength long, and shorted (from an RF standpoint) at one end will have a very high impedance at the other end. In other words, it acts like a parallel tuned circuit! We will use this characteristic later when we discuss microwave filters.

A quarter wavelength transmission line of any type (coax, open wire, stripline, microstrip, etc) can be used to transform impedances. This is called a "quarter wave transmission line transformer". The characteristic impedance of the quarter wave section should be set to the square root of the product of the input and output impedances. As an example, if it desired to connect a 50 ohm system to a 100 ohm system, it is only necessary to use a quarter wave length of transmission line *(remember the velocity factor)* having an impedance of $\sqrt{50 \cdot 100} = 70.7\,\text{ohms}$.

Transmission line **stubs** that are either open or shorted at the end have useful properties as electrical components. The stub will act as either a capacitor or an inductor, depending on its electrical length (expressed in degrees, where 360 degrees corresponds to one wavelength on the transmission line). The expressions for determining the actual value of inductive or capacitive reactance for stubs of less than a quarter wavelength are as follows:

For shorted stubs, $X_L = Z_0 \tan\theta$ or $\theta = \tan^{-1}\left(\frac{X_L}{Z_0}\right)$

For open stubs, $X_C = \frac{Z_0}{\tan\theta}$ or $\theta = \tan^{-1}\left(\frac{Z_0}{X_C}\right)$

where θ is the electrical length of the stub in degrees.

Note that it is also possible to determine the length of a shorted or open stub directly from a Smith Chart (described in a later chapter).

Power Splitters And Couplers

When splitting a signal into two paths, or when combining two signals to one, devices called **power splitters** and **power combiners** are used. These are designed so that the two paths have minimal interaction between them. Using microstrip, a common circuit is the Wilkinson power divider, as shown below:

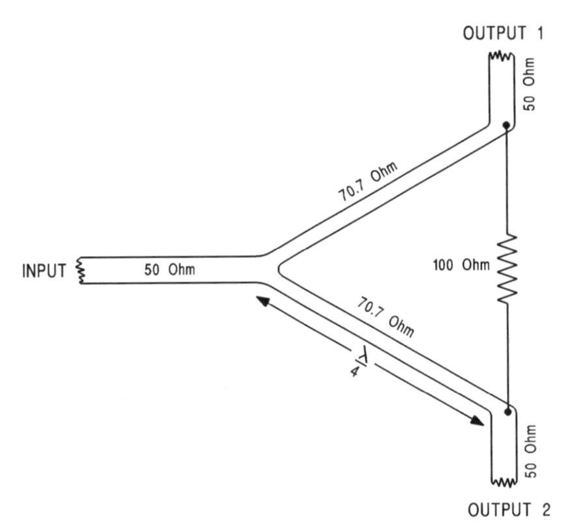

No power is dissipated in the 100 ohm resistor if the two loads are equal. The two outputs of a Wilkinson power divider are in phase, and each is 3 dB below the input power. Because the design of this power divider uses quarter wave sections of microwave transmission line, it is necessarily frequency sensitive. At frequencies other than the design frequency, the quarter wave lines will appear either too short or too long. It is useful to characterize this three port power divider as a function of frequency relative to the design centre frequency. This is done by normalizing frequencies - 1.0 represents the designed centre frequency, and other frequencies are represented as a ratio of it. We will plot the input and output VSWR (defined in a subsequent chapter), and the isolation between the two output ports as a function of the normalized frequency:

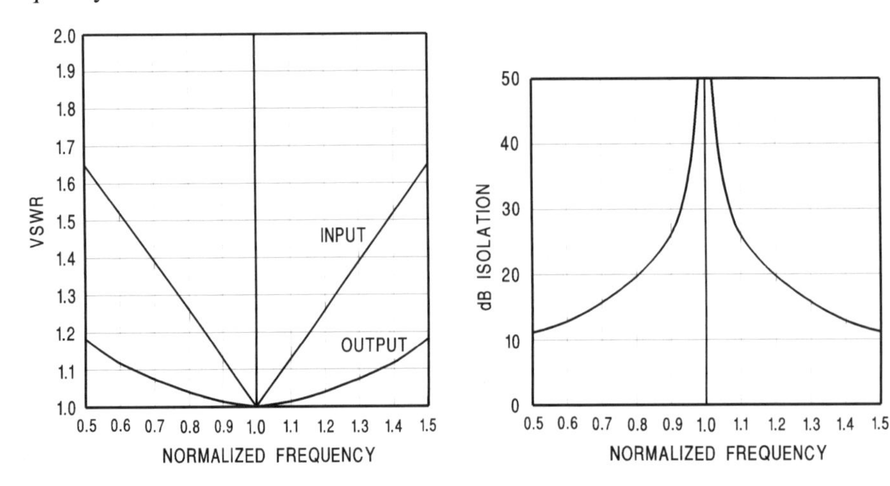

A special type of coupler called the **branch line coupler** is often used at microwave frequencies as a separate device, or as part of a mixer. It has four ports, and consists of four quarter wave sections of microstrip transmission line. The ports are labeled as A, B, C, and D. From each of the ports, microstrip at the system impedance Z_0 (usually 50 ohms) connects to the respective corner of the square formed by the two shunt elements and the two series elements. The impedance of the shunt microstrip elements is different from the impedance of the series microstrip elements:

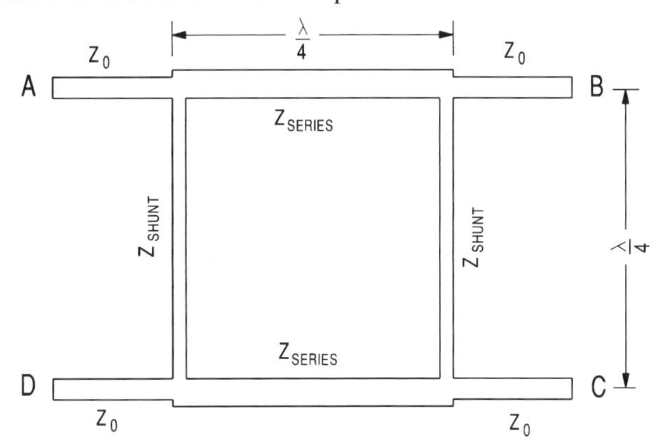

Power fed into port A is split to appear at ports B and C. Power fed into port D likewise splits into ports B and C. The coupling between the lines AB and DC can be adjusted by changing the relative impedances of the shunt and series arms. In order to achieve 3 dB coupling (in other words, the power is equally split), the impedances of the two series arms should be set to $1/\sqrt{2}$ of the shunt impedances. Therefore, for a 50 ohm system, the impedance of the series arms should be 35.4 ohms, and the shunt arms should be set to 50 ohms.

Power fed into port A of a 3dB branch coupler will split equally between ports B and C, but none will flow to port D if the two output loads are equal. Power fed into port D will also flow equally to ports B and C, but none will flow to port A. This means that there is isolation between ports A and D. The isolation is easy to understand if you consider a signal entering in port A, and then add up the phase delays as it winds its way to port D - it will be found that there are two out of phase signals which cancel at port D.

The following charts show the performance of the branch line coupler:

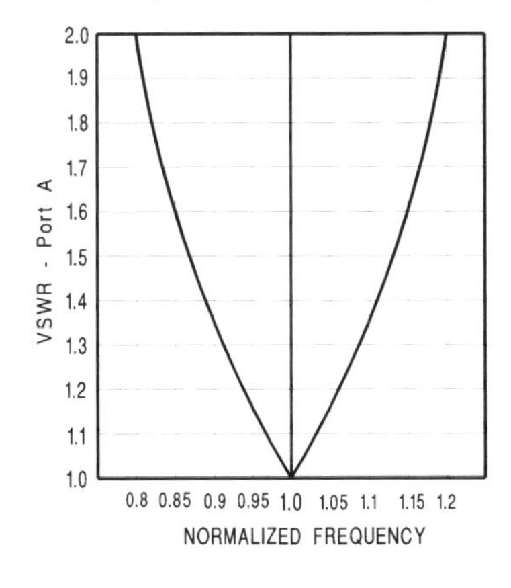

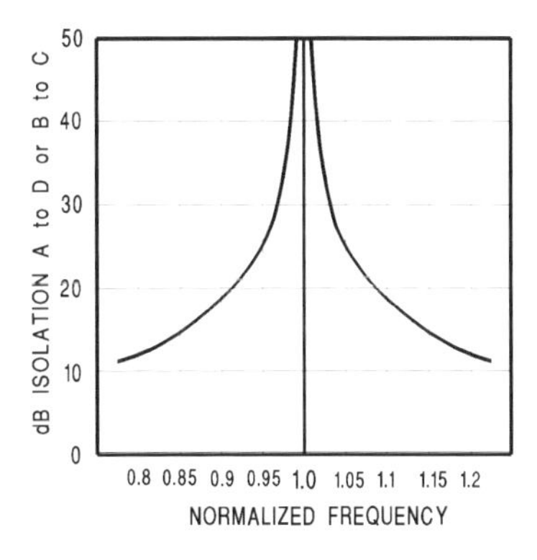

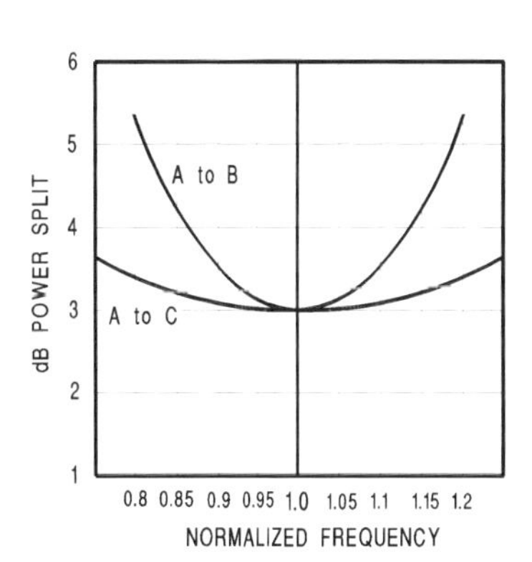

This component is normally used with a 50 ohm terminator on one of the input ports (either A or D), and the input signal on the other. The two outputs are 90 degrees out of phase, and are taken from ports B and C. No power is dissipated in the terminator if the two loads are equal. This device could of course be built using stripline instead of microstrip techniques. Here is an example mixer circuit using a branch line coupler. Note that after the two diodes, there are two open-circuit "stubs" which are a quarter wave long at the RF and LO frequencies respectively, thereby providing a low impedance.

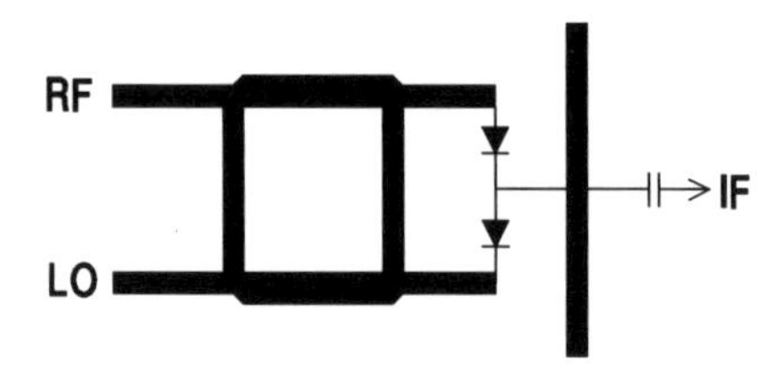

A very interesting component is the so-called **rat-race hybrid** or **magic tee**. It consists of a ring that is 1.5 wavelengths in circumference with four arms that are separated by 60 degrees of angular rotation. Note that this device does not actually need to be circular in shape, just as long as the correct electrical lengths are used. The impedance of each of the four arms (A, B, C, and D) is set equal to the system impedance Z_0 (usually 50 ohms), while the impedance of the ring is set equal to $\sqrt{2}$ times Z_0.

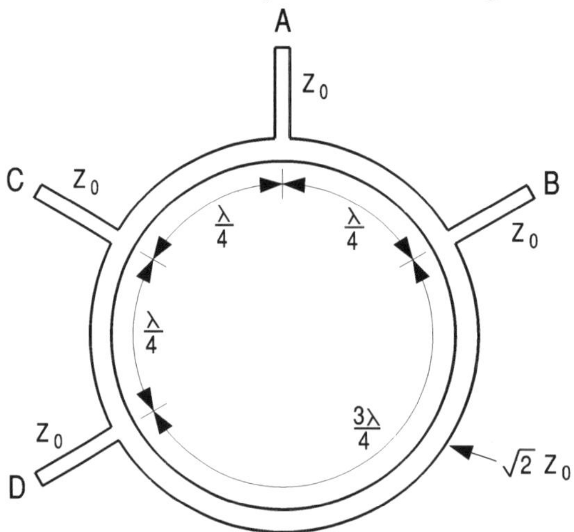

If power is fed into port A, it will be split equally to appear at ports B and C with similar phases. If port D is terminated in the characteristic impedance, there will be good isolation between the two output ports. Alternatively, the input can be applied at port D, a terminator placed at port A, and two equal but opposite phase signals taken from ports B and C. These two configurations are shown below:

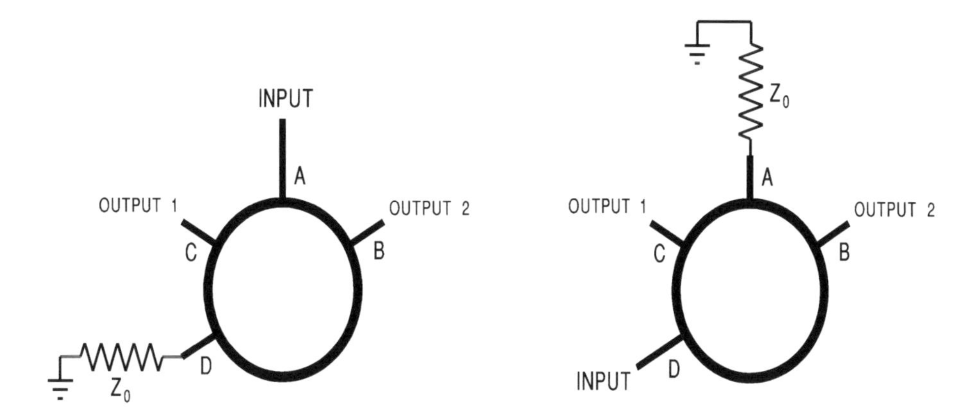

The impedance of all sections is set equal to $\sqrt{2}$ times the port impedances. In a 50 ohm system, the impedance of the ring should therefore be set to 70.7 ohms. The performance of the rat-race hybrid is shown below:

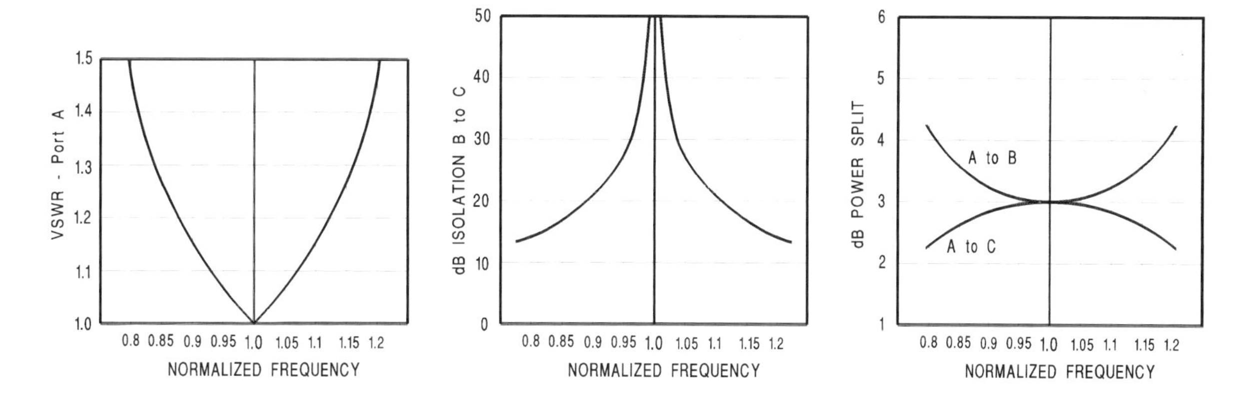

Both branch line couplers and rat-race hybrids are commonly used in singly balanced mixers. The RF signal is fed into the main port, and the LO is fed into the port that was described above as being terminated. The two outputs (either 90 or 180 degrees out of phase) are connected to two series connected mixer diodes. The IF is taken from the junction of the two diodes.

Microstrip Directional Couplers

Directional couplers can easily be fabricated with microstrip. Two microstrip line segments that are in close proximity to each other will exhibit a directional coupling path. The degree of coupling is a function of the spacing between the two lines.

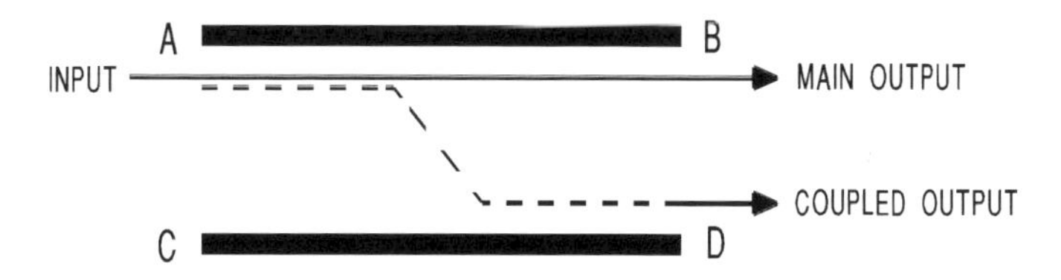

Power entering the coupler at port A mostly flows straight through to port B, but some portion of it is split off by the coupling mechanism to port D. A small amount of undesired power also appears at port C - its value is a function of the coupler's directivity.

High coupling and directivity can be obtained if the microstrip segments that are coupled together are one quarter wavelength long. In order to design couplers having specific responses, it is necessary to understand odd mode and even mode impedances, which are abbreviated as $\mathbf{Z_{oo}}$ and $\mathbf{Z_{oe}}$. These terms refer to impedances of parallel microstrip conductors which are in near proximity.

The **even mode** characteristic impedance Z_{oe} is the impedance of one of the conductors to ground when the other conductor is at the same potential, and is passing current in the same direction. The **odd mode** characteristic impedance Z_{oo} is the impedance of either conductor to ground when the other conductor is at the opposite potential, and carrying current in the opposite direction.

If the desired voltage coupling factor is "c", the necessary odd and even mode impedances can be calculated as:

$$Z_{oe} = Z_o\sqrt{\frac{1+c}{1-c}} \qquad Z_{oo} = Z_o\sqrt{\frac{1-c}{1+c}}$$

Looking at these equations, it can be seen that for any coupling factor,. It is possible to calculate Z_{oo} and Z_{oe} on the basis of the dimensions of the conductors, but the math is quite complex. Assuming that the nominal impedance of the system is 50 ohms, the following table gives some idea of the dimensions to use using common PCB materials:

			ε = 4.8				ε = 2.2			
			1/16 Fiberglass		1/32 Fiberglass		1/16 Teflon		1/32 Teflon	
Coupling	Z_{oo}	Z_{oe}	W	s	W	s	W	s	W	s
6 dB	29	87	67	6.4	32	4.4				
10 dB	36	69	92	15	44	8.1	156	12		
15 dB	42	59	105	42	51	21	181	34	88	18
20 dB	45	55	109	63	53	41	189	79	92	39
25 dB	47	53	110	135	54	67	191	141	93	70
30 dB	48	51	111	206	54	102	192	224	94	112

Where W is the width (in mils) of the coupled sections, and s is the separation (in mils). A coupler can be created using this information as follows:

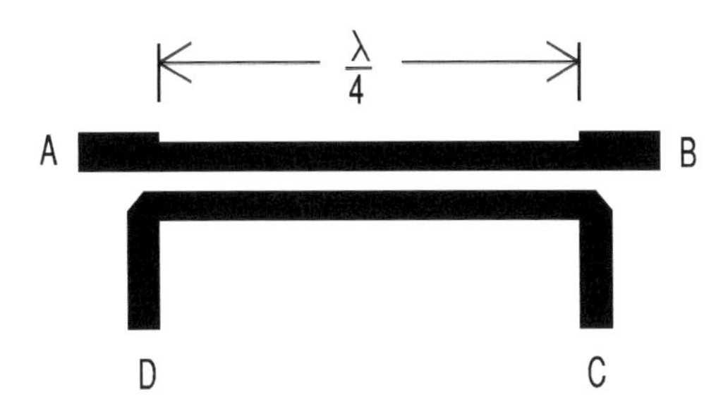

Note that if port C is terminated, then port D will have an output that is less than the input on port A by the coupling ratio of the structure, and it will respond to forward power only. If port D is terminated, port C will be proportional to reflected power from the load. This is a useful component that can be used to determine how well a load is matched.

Microstrip Filters

At frequencies over 1 GHz, filters are easier to fabricate using transmission line sections rather than discrete components. The filters can be included as part of a PCB layout, thereby reducing cost and increasing reproducibility. Either microstrip or stripline is commonly used.

Bandpass filters are made by using coupled transmission line segments: either short-circuited quarter wave segments, or open-circuited half wave segments. Two typical bandpass filters are shown below. Note that the one on the right simply uses "folded over" half wave sections to save space: it is called a "hairpin" filter:

Two other passband filters that are often used in microwave receiver design are the comb-line filter and the interdigital filter. The following two design examples are drawn as though the filters are fabricated from separate mechanical parts (indeed, they could very well be implemented using machined lines), but they could just as easily be etched into microstrip. Adjustment capacitors are shown so that the design can be centered precisely on the frequency of interest. As a simplification, consider that these designs are just high frequency equivalents to the coupled LC filter designs that were discussed in an earlier chapter. Note that the input and output connections are “tapped down” the end elements - this is analogous to the method discussed earlier of transforming the load impedance by tapping down the inductor.

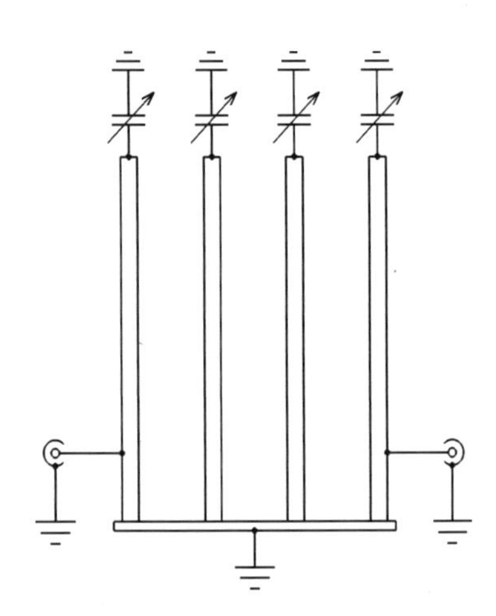

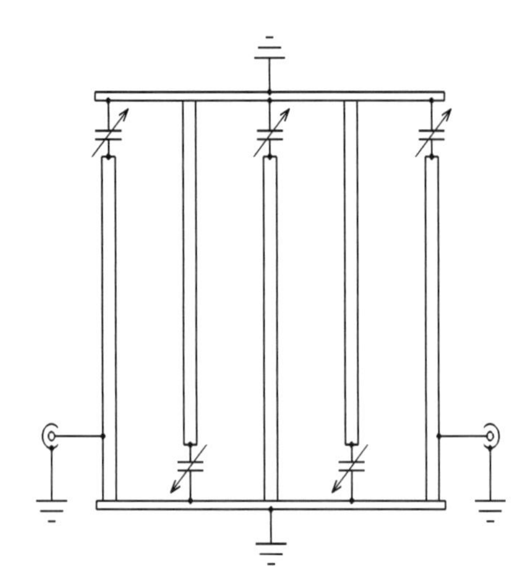

In all of these coupled structure filters it is possible to synthesize all the common filter types and shape factors (Butterworth, Elliptic, Chebyshev, etc) by adjusting the section lengths and the coupling factors. Designing microstrip filters can get quite involved! Fortunately, software is available to help with this process. Sonnet Lite is a commercial software package that can be downloaded at no cost from sonnetsoftware.com. It allows microstrip circuits to be designed and/or analyzed in the frequency range of 1 MHz to many GHz.

ATTENUATORS

An attenuator is a device or circuit that reduces the amplitude of a signal. You have already been exposed to one basic form of an attenuator - the classical two resistor voltage divider:

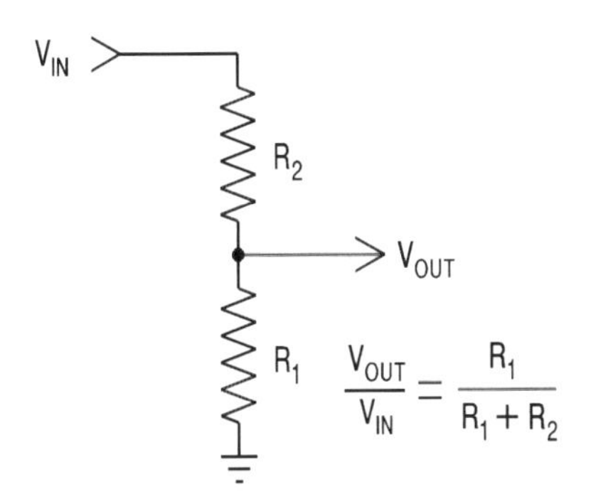

This simple resistive divider works just fine for DC and audio frequencies, but is not an ideal configuration at RF. The two resistors have a small amount of series inductance in their physical structure, as well as capacitance from end to end and to ground. The net effect is that the voltage division ratio will be frequency dependent. In order to minimize this, very small value resistors can be used, but this might unduly load the source. Another problem with this divider is that the input and output impedances are not equal. A two resistor voltage divider can not be put into a transmission line without causing a mismatched condition.

In order to maintain a constant impedance (usually 50 ohms), three resistors are needed for an RF attenuator. The circuit configuration can be either a "T" or "Pi" arrangement:

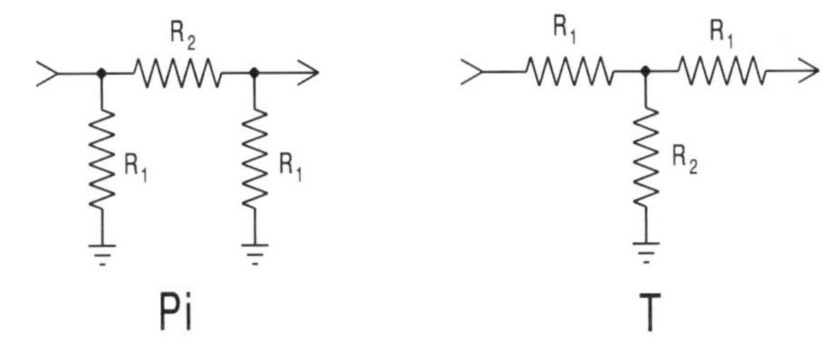

By choosing appropriate values, any attenuation ratio and system impedance can be implemented. Note that these attenuators are symmetrical, and can be used in either direction. Resistor values (in ohms) for a few common attenuation levels in a 50 ohm system are as follows:

ATTENUATION	Pi Network		T Network	
dB	R_1	R_2	R_1	R_2
1	870	5.8	2.9	433
3	292	17.6	8.5	142
6	151	37	16.6	66.9
10	96.2	71.2	26	35
20	61	248	41	10
30	53	790	47	3.2
40	51	2500	49	1

If the attenuator is going to be attenuating any appreciable power, significant heat will be generated in the attenuator. Commercial attenuators all have a specified maximum power handling capability.

It is sometimes desirable in system design to have an RF attenuator that can be controlled by a DC voltage. Applications would include AGC systems, and transmitter power control. If the frequency is less than 1 MHz, a FET can often be used as a controllable resistor if the signal peak-to-peak amplitude between source and drain is less than a few hundred millivolts. For higher frequencies, the PIN diode is the device of choice.

A PIN diode behaves like a fairly conventional diode for DC and audio frequencies, but it acts like a controllable resistor for RF frequencies. The lowest frequency that is usable is a function of the device's carrier lifetime, which is specified by the manufacturer. Most PIN diodes are usable down to a few Megahertz. The effective RF resistance is a function of the DC current through the diode. A typical set of RF resistance data for a low power PIN diode is as follows:

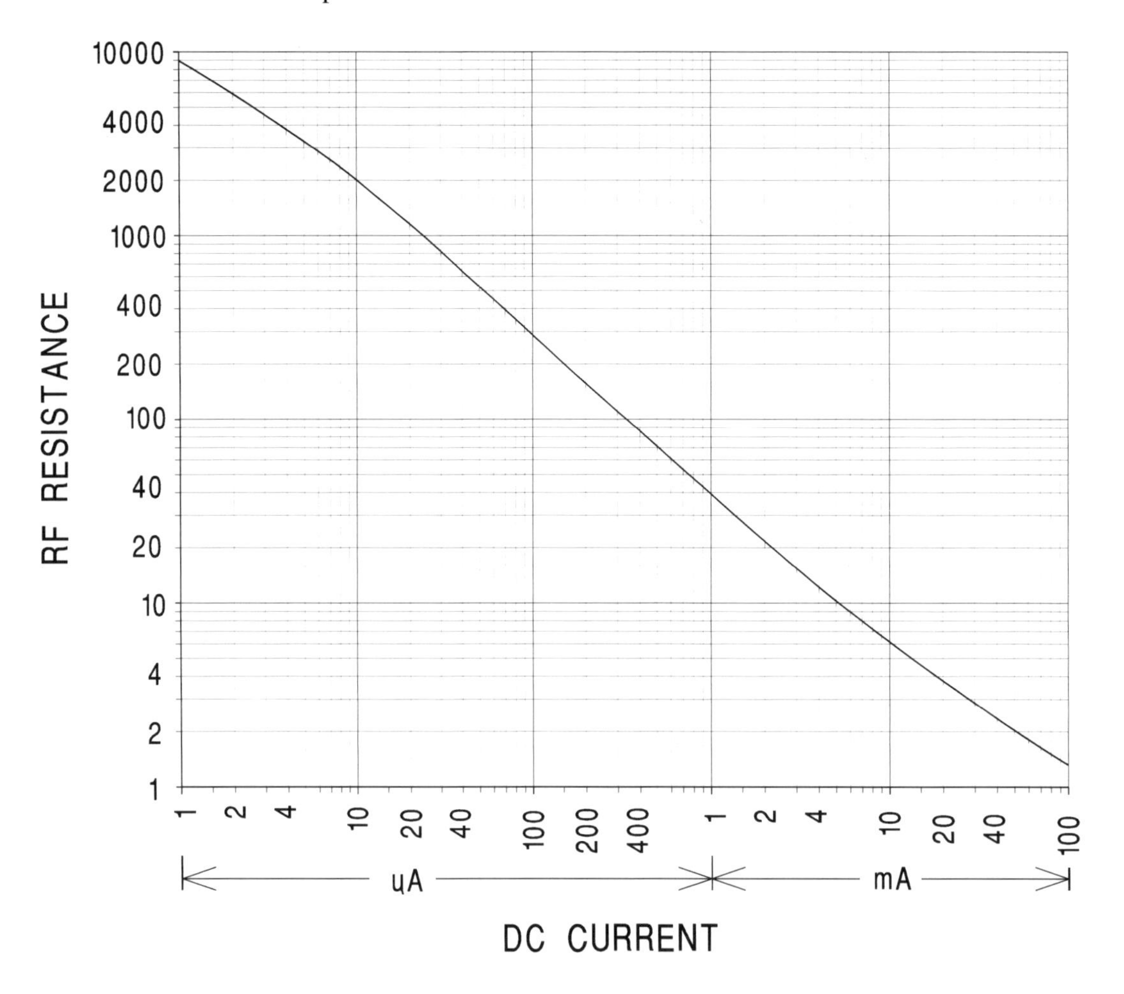

By careful design, it is possible to create either T or Pi attenuators using three PIN diodes that work over a wide attenuation range while maintaining a fairly close match to a 50 ohm source and load impedance.

REFLECTIONS

As mentioned before, transmission lines are used to transfer power from one device (the "source") to another device (the "load"). If an infinitely long transmission line is connected to a source, the power never gets to the other end, and the impedance seen looking into the transmission line is equal to Z_0, the line's characteristic impedance. In the real world, infinitely long transmission lines are not used, because we would like to get power to the load at the other end. If an infinitely long transmission line is cut at some point (perhaps 1 metre away from the source) and a resistor whose resistance is equal to Z_0 is connected instead of the rest of the infinite line, the source will not be able to tell the difference - it still sees Z_0 as a load, and has no idea if the power ever gets to the other end. In other words, we can "fool" a source into thinking that a line is infinitely long by using a finite length that is terminated in its characteristic impedance.

If the source impedance is equal to the characteristic impedance of the transmission line and is also equal to the load impedance, all of the power travels smoothly from source to load without any reflections. Unfortunately, the impedances often are not all equal, and reflections will occur. Consider the following diagram:

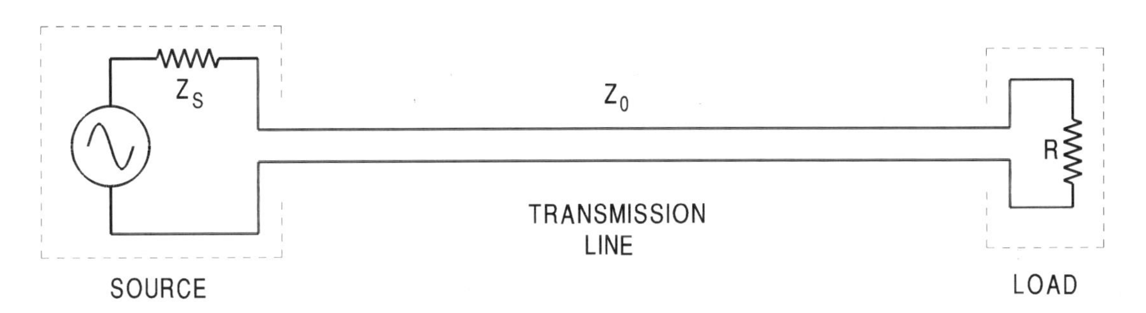

If ZS = Zo = R, then the peak voltage and current is the same throughout the transmission line, and all of the energy travels from left to right. If R is not equal to Z0, the load doesn't look like "more of the transmission line", and some power is reflected from the load. If R is set to 0 (a short circuit), all of the power is reflected back. Because R is zero, there can not be any voltage developed across it, so the voltage of the reflected signal must be 180 degrees out of phase with the incident signal in order to cancel completely. The reflected current is in phase with the incident current at the load.

If the load is an open circuit, no current can flow in it, therefore the reflected current must be the same as the incident current, but 180 degrees out of phase with it. The reflected voltage is in phase with the incident voltage.

If R is somewhere between a short and an open, but it is not equal to Zo, only some portion of the incident wave is reflected.

The ratio of the reflected voltage to the incident voltage is called the reflection coefficient:

$$\Gamma = \frac{E_r}{E_f}$$

where Γ = reflection coefficient (pronounced "Gamma")
E_r = reflected voltage
E_f = forward voltage

Note that many amateur radio publications use the letter ρ in place of Γ to represent the reflection coefficient.

If Γ is equal to zero, it indicates that all of the power is delivered to the load, and none is reflected. If Γ is 1, it indicates that all of the power is reflected.

For the moment we will make the simplifying assumption that the load is always resistive, and has no reactive component. Then,

$$\Gamma = \frac{R - Z_o}{R + Z_o}$$

If Γ is positive, it indicates that R is greater than Z_o. If Γ is negative, R is less than Z_o. We have been dealing with resistive loads up to now, but this is seldom the case. Remember that Γ is actually a complex quantity, having both magnitude and angle.

In order to get a better "feel" for the issue of reflections, we will start out by examining some examples based on DC transients.

Consider a switched DC voltage source having a source resistance of 100 ohms that is connected to a transmission line with a characteristic impedance of 100 ohms. The transmission line is 10 feet long, and it is connected to a load resistor of 50 ohms. Because of the speed of light, it takes approximately 10 ns for a signal to go from one end of the transmission line to the other. The source impedance is equal to the transmission line's characteristic impedance, so the reflection coefficient of the source is zero. At the load however, there is a miss-match, and the reflection coefficient at the load can be calculated as:

$$\Gamma_L = \frac{R - Z_o}{R + Z_o} = \frac{50 - 100}{50 + 100} = -0.333$$

The circuit looks like this:

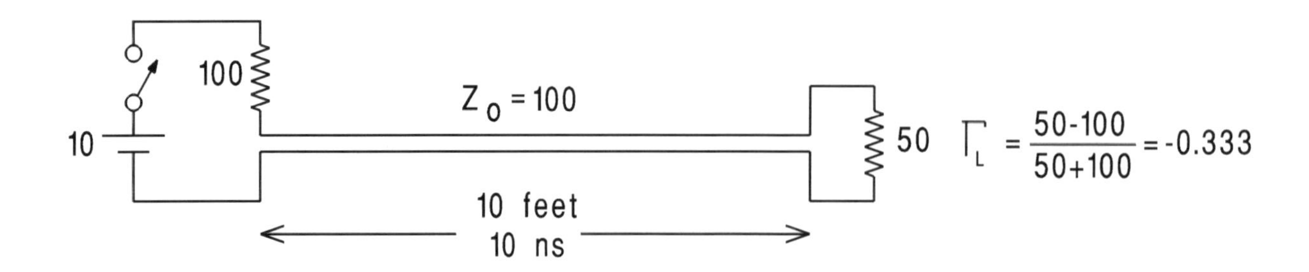

At time T=0, we close the switch. The instantaneous impedance looking into the transmission line at T=0 appears to be 100 ohms, so a voltage divider is formed between the source impedance and the line impedance, and therefore a **5 volt** level starts propagating down the line. Note that for the first few nanoseconds, the source is not even aware that there is a load resistor at the other end of the transmission line. After 5 ns, the 5 volt level has propagated half way down the transmission line, and the voltage distribution along the line looks like the following diagram:

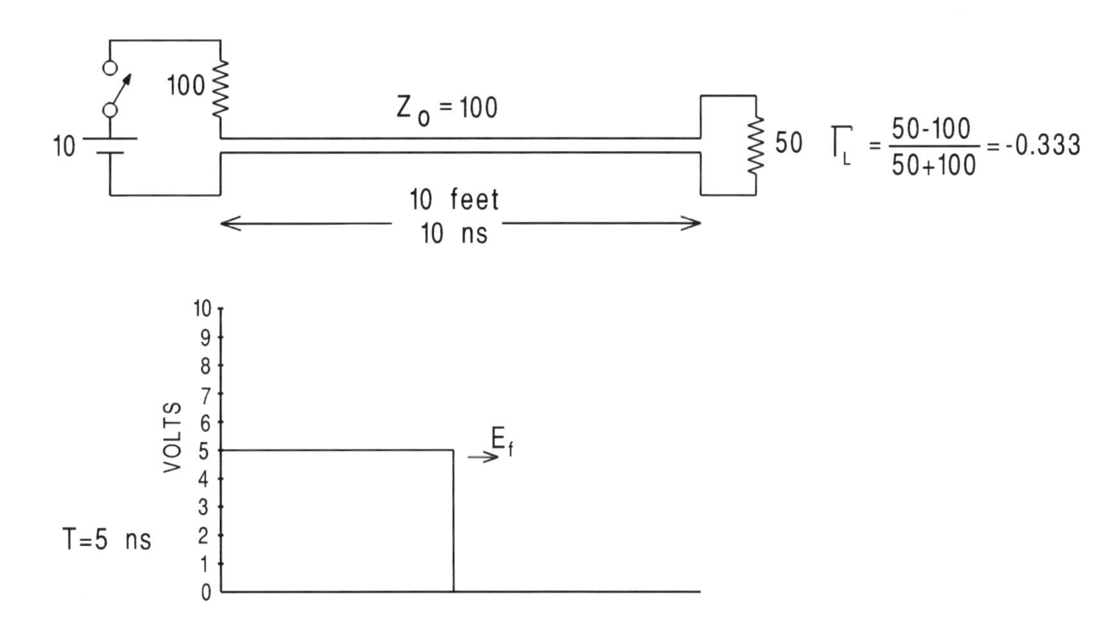

After 10 ns, the 5 volt level will reach the load, and a portion of it will be reflected back toward the source. The reflected voltage is equal to the incident voltage times the reflection coefficient, or 5 x -0.333 = -1.666 volts. This reflected signal will propagate back toward the source, and it will subtract from the 5 volt forward signal. After 11 nsec, the voltage distribution will look like:

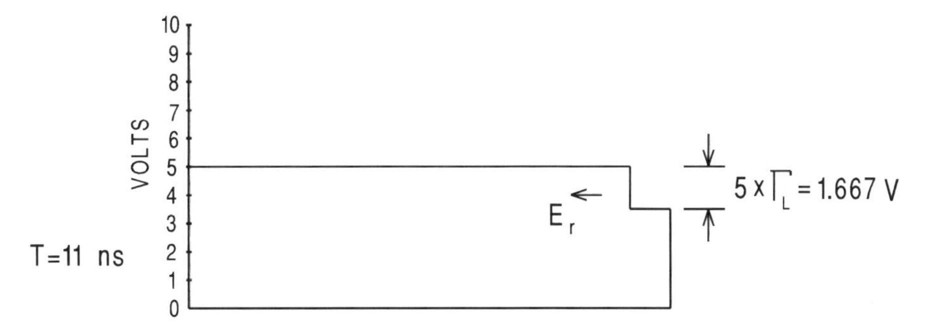

This -1.666 V reflected wave will continue to travel back toward the source. It will reach the source 20 ns after the switch was closed. At the source, it will not be reflected, because the source reflection coefficient is zero. Therefore no further reflections occur, and the entire line now has a steady state DC voltage of 3.333 from one end to the other. This is just what could have been predicted by looking at the ratio of the load and source resistors, but without worrying about the transient effects due to wave propagation and reflections.

We will now look at an example where both the source and the load are miss-matched to the transmission line. The following diagram traces the reflections for the first 24 ns:

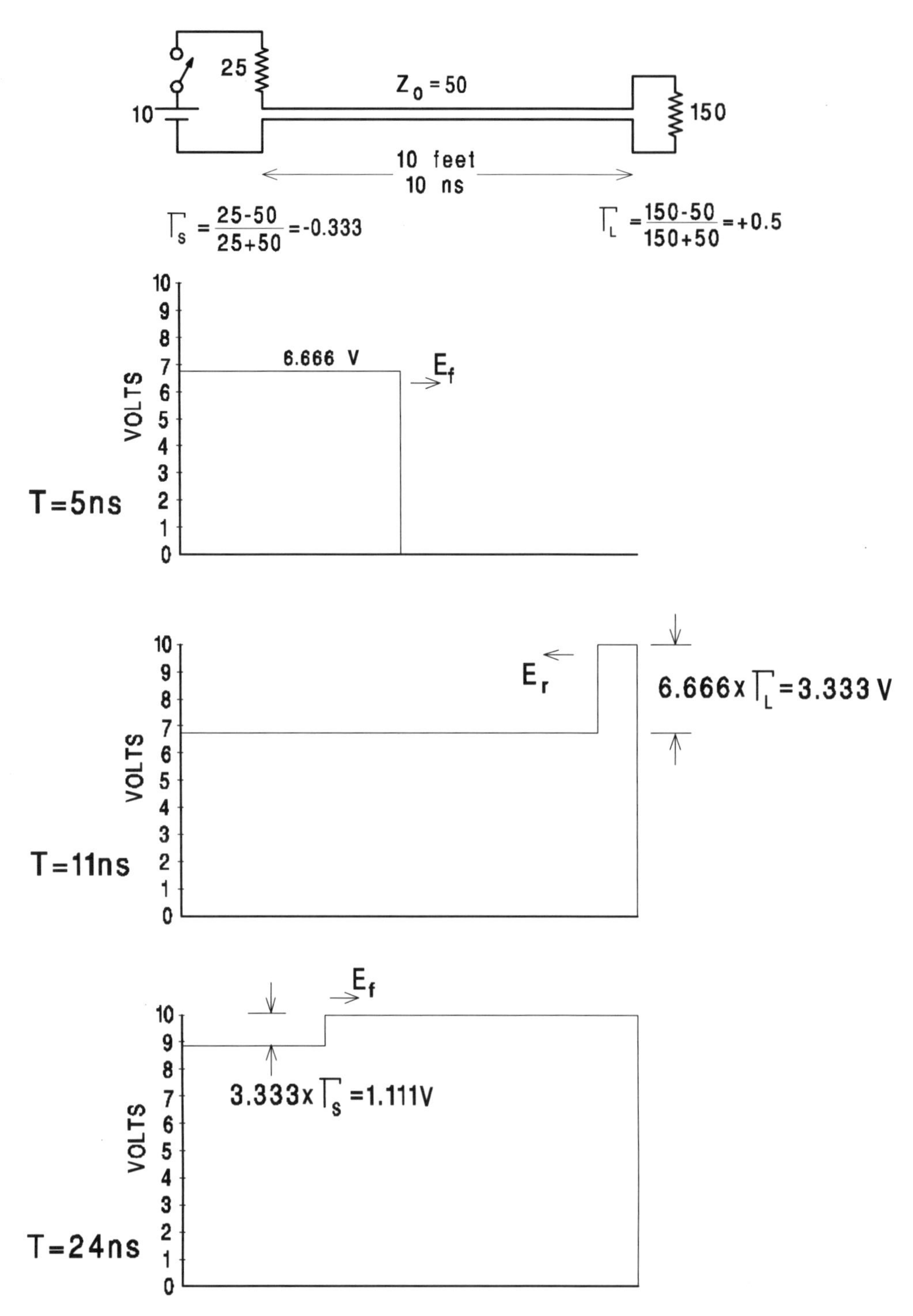

Note that neither the source nor the load resistances match the transmission line impedance, and therefore there is both a source reflection coefficient and a load reflection coefficient. It can be seen that the initial voltage wave that started flowing when the switch was closed will continue to bounce back and forth along the transmission line indefinitely, although the magnitude of the reflected signal is getting less at each reflection.

The following diagram plots the voltage across the load resistor as a function of time:

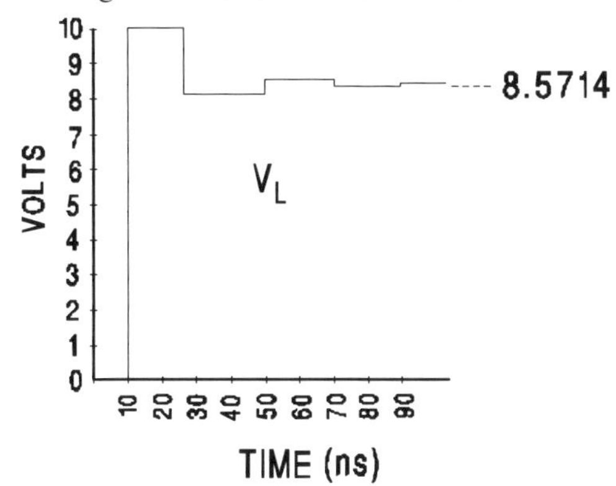

After 4 or 5 reflections, the voltage across the load settles out to approach its final value of 8.5714 volts that could be predicted just by looking at the resistor ratios. Even though it is easy to predict the final voltage by just applying ohms law, it is obvious that a lot of activity takes place on the transmission line before equilibrium is approached!

The preceding two examples dealt with just a single event (the closing of the switch). We will now turn our attention to AC waves, where the source continues to deliver alternating voltages as the forward and reverse signals propagate up and down the transmission line. Because of the interaction of the incident and reflective waves, their constructive and destructive interference will result in standing waves occurring on the transmission line. The electrical distance between voltage peaks or nulls is 180 degrees. If the system is perfectly matched, the voltage will be constant throughout the length of the transmission line, but if the load resistance is not equal to Z_o, the effect of the reflected power is to interfere with the incident power and produce voltage peaks and nulls along the line.

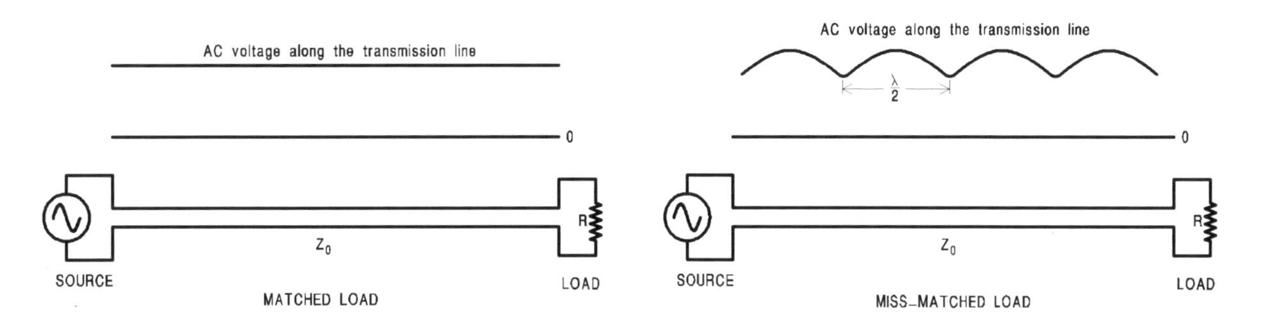

A slotted transmission line can be used with a moveable probe and RF voltmeter to measure the magnitude and location of the peak and minimum voltages. The ratio between these two values is called the Voltage Standing Wave Ratio (**VSWR**). Many people just call this ratio the SWR ratio, but it is much easier to say the word "VSWR" (pronounced *vis-w-are*).

If you know what the VSWR is, either by using a slotted line or a directional coupler, you can determine the magnitude of the reflection coefficient Γ:

$$|\Gamma| = \frac{\text{VSWR} - 1}{\text{VSWR} + 1} \quad \text{also,} \quad \text{VSWR} = \frac{1+\Gamma}{1-\Gamma}.$$

Note that VSWR is just a numerical ratio, while Γ has both magnitude and angle.

If the load has both resistance and reactance, it is still possible to determine the VSWR:

$$\text{VSWR} = \frac{A+B}{A-B}$$

where: $A = \sqrt{(R+Z_o)^2 + X^2}$

$B = \sqrt{(R-Z_o)^2 + X^2}$

R = the resistive component of the load

X = the reactive component of the load

The maximum voltage that will be present on a transmission line increases with VSWR:

$$E_{MAX} = \sqrt{P \cdot Z_o \cdot \text{VSWR}}$$

where: P = power delivered by the source

E_{MAX} = peak voltage on the line

This last equation indicates that the peak AC voltage along the transmission line increases as the VSWR increases. Even moderate amounts of power could conceivably cause voltage breakdown in the transmission line if the VSWR is very high due to a grossly miss-matched condition (such as a short or open in the line).

Transmission lines have losses at the operating frequency that are measured in dB per unit length. The effect of these losses is to reduce the VSWR as seen by the source. The VSWR is highest at the load, then decreases along the line back to the source because of attenuation of the reflected wave. This is because the line losses are attenuating the reflected power as well as the incident power.

If the load is not purely resistive, Γ can be calculated as follows:

$$\Gamma = \frac{Z_L - Z_0}{Z_L + Z_0}$$

where Γ = the reflection coefficient (magnitude and angle)

Z_L = the complex load impedance

Z_0 = the system impedance

For resistive loads, VSWR is simply the ratio of the load or source resistance to Z_0.

VSWR can be measured by using a directional wattmeter which separately measures the forward and reverse power on a transmission line:

$$\text{VSWR} = \frac{1+\sqrt{\dfrac{\text{refl pwr.}}{\text{fwd pwr.}}}}{1-\sqrt{\dfrac{\text{refl pwr.}}{\text{fwd pwr.}}}}$$

The relationship between the VSWR and the ratio of forward to reverse power can also be expressed as:

$$\text{Power Ratio (fwd/rev)} = \left(\frac{\text{VSWR}-1}{\text{VSWR}+1}\right)^2$$

When using a directional wattmeter or directional coupler for low power levels, it is common to specify the fraction of reflected power to the incident power in dB. In this case, the dB power ratio is referred to as the "Return Loss". The relationship between VSWR, power ratio, and return loss is shown in the following table:

VSWR	POWER RATIO (FWD / REV)	RETURN LOSS (dB)
1.00	0	infinity
1.02	0.0001	40
1.07	0.001	30
1.22	0.01	20
1.93	0.1	10
3.00	0.25	6
3.57	0.316	5
5.84	0.5	3
infinity	1.0	0

The mathematical relationship between VSWR and return loss is:

$$\text{Return Loss (dB)} = 10\log_{10}\Gamma^2 \qquad \text{VSWR} = \frac{1+10^{\frac{-RL}{20}}}{1-10^{\frac{-RL}{20}}}$$

We will use many of the concepts discussed here regarding reflection coefficient in the next section that discusses S-parameters.

SCATTERING PARAMETERS

Many RF circuits or components can best be analyzed by considering them as "two port networks". The object under analysis is treated as a "black box", and measurements are made at its two ports. As an example, a transistor amplifier can be considered as having an input port and an output port, and it can be characterized based on measurements made on these ports under controlled conditions. A filter can also be considered as a two port network.

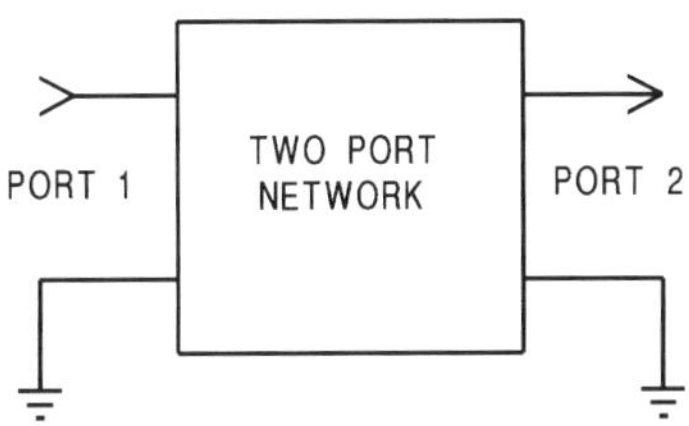

This "two port approach" is very convenient at RF frequencies, and it is commonly employed in radio design work. Many different schemes have been developed over the years to characterize a two port network: Y-parameters, h-parameters, ABCD parameters, etc. The most convenient and widely used approach in RF work is called "Scattering Parameters", or "S-Parameters" for short. The use of S-Parameters is so popular because it is fairly easy to measure S-parameters using automatic equipment, and most RF active components include this data on their specification sheet.

There are four S-Parameters, labeled S_{11}, S_{12}, S_{21}, and S_{22}. They are based on the measurement of reflection or transmission coefficients at the device's two ports. An S-parameter is a voltage ratio, and has both magnitude and angle. S-Parameters are measured with the ports terminated in a characteristic reference impedance, which is most often 50 ohms.

S-parameters are a function of frequency, and a specification sheet will give these parameters at a number of frequencies, or may provide a graph. S-Parameters are measured using a signal source, directional couplers, and vector voltmeters. S_{11} is the network input reflection coefficient, and it is measured somewhat like you would measure the VSWR into an antenna except that it is necessary to also know the phase angle of the reflected signal (if any). A signal source is connected via the directional coupler into the input port of the device to be tested, while the output port is terminated in 50 ohms. S_{11} is defined as the input reflection coefficient (Γ) when the output port is terminated in the system impedance (usually 50 ohms), and it has both magnitude and angle. S_{11} is the ratio of the incident wave to the reflected wave on port 1. The following sketch illustrates an experimental setup for measuring S_{11}.

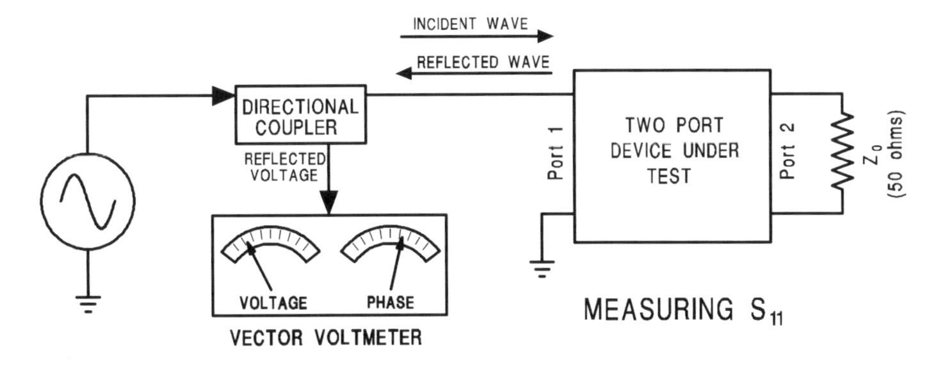

The magnitude of S_{11} is either specified as a number less than 1, or in dB (with a negative sign). As an example, if the input port of a network consisted of a pure 50 ohm resistance, then S_{11} would be zero. If the input port looked like a dead short circuit, then S_{11} would be 1.0, with an angle of -180 degrees. If the input port looked like an open circuit, then S_{11} would be 1.0 with an angle of zero.

Similarly, S_{22} is the network's output reflection coefficient, and it is measure by connecting a signal source and directional coupler to the network's output port while the input port is terminated in the characteristic impedance (usually 50 ohms).

S_{21} is the forward transmission coefficient, and S_{12} is the reverse transmission coefficient.. This is equivalent to saying that these parameters are the network's forward gain, and the reverse gain. Note that both of these parameters again have both a magnitude and an angle. The reverse gain will normally be quite a small number, but it cannot be ignored.

For an amplifier, the magnitude of both S_{11} and S_{22} must be less than one. If either is greater than unity, we most likely have an oscillator!

It is possible to purchase expensive lab instruments called Vector Network Analyzers (VNA) that will precisely measure and plot S parameters as a function of frequency for one and two port RF devices. It is also possible to purchase relatively inexpensive (under $1000) small VNAs that connect to (and are powered by) a USB port on an attached laptop or Android device.

S parameters are a useful way of characterizing RF components, especially transistors and FETs. Data is commonly published in this form, and equipment is available that can automatically measure and plot this data as a function of frequency, although it tends to be rather expensive.

Stability is a messy issue with many RF amplifiers. Depending on the tuning of the input and output networks, RF amplifiers will often break out into unwanted oscillation. It is possible to use S-Parameters to determine if a particular device will be stable. For those who really want to know in advance if an amplifier will be "tricky" to tame, a stability factor K is defined as follows:

$$K = \frac{1 - |S_{11}|^2 - |S_{22}|^2 + |(S_{11}S_{22}) - (S_{12}S_{21})|^2}{2|S_{12}||S_{21}|}$$

This is called the "**Rollett Stability Factor**". If K > 1 and both S_{11} and S_{22} are less than 1, then the network is unconditionally stable. This means that the network will be stable for all input and output loads. If K is less than 1, there is the potential for oscillation at certain combination of load and source impedances.

S parameters are ideally suited for designing with Smith Charts, so we will need to discuss that topic next.

THE SMITH CHART

The Smith Chart is an extremely useful tool for analyzing RF networks. It was invented and published by Phillip Smith in 1939, and is still taught as a "modern" technique. Smith Charts are used to plot complex impedance or admittance, reflection coefficients, and determine the effect of lengths of transmission line.

The Smith chart is simply a chart of resistance and reactance that has been printed in a special circular manner. There are actually two types of Smith Charts: the **impedance chart** and the **admittance chart**. The two charts look the same, but are reversed 180 degrees from each other. We will start by looking just at the impedance chart. The chart consists of two groups of circles: one for resistance, and one for reactance. They are shown separately below:

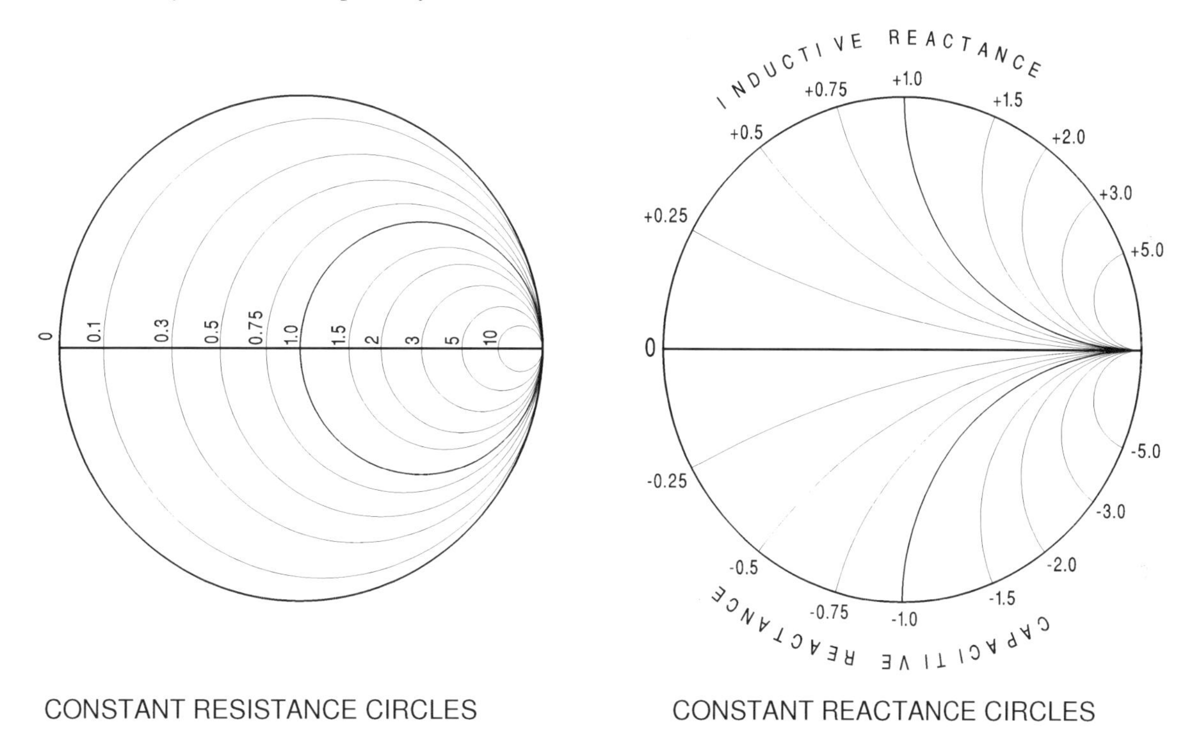

CONSTANT RESISTANCE CIRCLES

CONSTANT REACTANCE CIRCLES

A Smith Chart has both sets of circles superimposed on top of each other. Any point placed on the chart will therefore have a corresponding value of resistance and reactance (positive or negative).

The following complete chart plots two impedances: 0.5 + j1 ohms, and 1 + j2 ohms. Since both of these impedances are inductive, they both appear in the upper half of the chart.

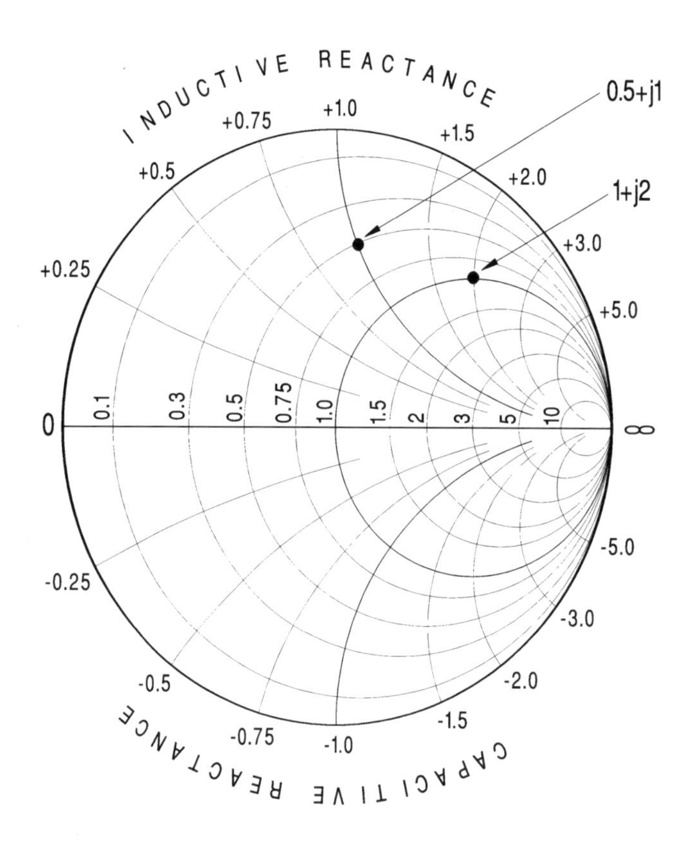

Note that the centre point of this chart represents 1 + j0 ohms. This is quite a low value of impedance, so it is usual to **normalize** impedances to the characteristic impedance of the system (usually 50 ohms). This means that all impedances are divided by Z_0 before plotting. Therefore, the centre of the normalized chart will usually represent a pure resistance of 50 ohms. A full-sized Smith Chart also has a set of **radially scaled** parameters that are drawn on the bottom. A complete Smith Chart is shown on a separate page in this section.

One property of this chart is that all impedances representing a given VSWR will appear on a circle whose centre is at the prime centre of the chart. The value of the VSWR can be read directly off of the right hand real axis. As an example, draw a circle centred on the prime centre, and passing through the point 2.5 + j0. Now, every impedance on this circle represents a load having a VSWR of 2.5. As an example, the following load impedances are all found on this circle, and all have a 2.5 VSWR (note that these have been normalized for a 50 ohm system): (125 + j0), (20 + j0), (30 + j30), (60 - j50). We have already found one use for the Smith Chart: converting load impedances to VSWR!

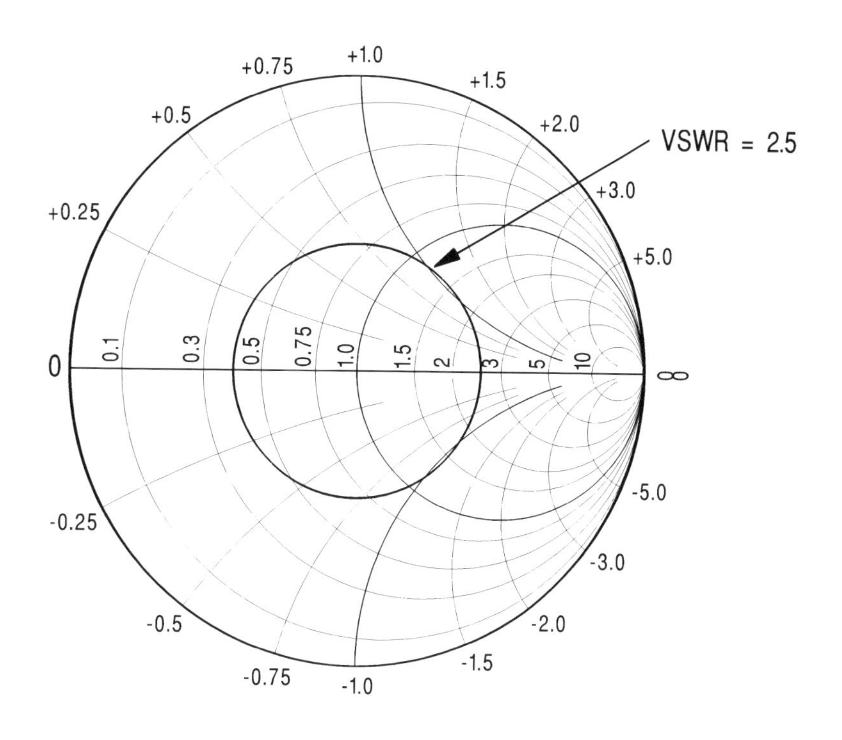

On the outside of a complete Smith Chart is a set of graduations representing distance along a transmission line. The units are expressed in terms of fractions of a wavelength. Clockwise motion represents movement along a transmission cable toward the source, and counter-clockwise rotation represents movement toward the load.

Since all points on a circle centred on the prime centre of the Smith Chart have the same VSWR, we can see the effect of introducing a length of transmission line into a system. We will ignore cable losses in this discussion. Assume that our load has an impedance of (20 + j0) ohms. We draw a circle centred on the prime centre that passes through this point, observing that the VSWR is 2.5. Now let's connect a length of 50 ohm transmission line to this load and try to determine what impedance will be seen at the opposite end of the line. At our frequency of interest, let's for the moment assume that the electrical length of the line (including velocity factor) is 0.15 wavelengths. Look at the outside rim of the chart and find the circular scale labeled "wavelengths toward generator", and follow it clockwise around to the 0.15 wavelength point. Now draw a straight line from the chart's prime centre to the 0.15 wavelength point, and note that it intersects our 2.5 VSWR circle at an impedance of (45 + j45) ohms. This is the impedance that we would see looking into the line. If the line were 0.25 wavelengths long, you can easily see that the impedance would be (125 + j0) ohms.

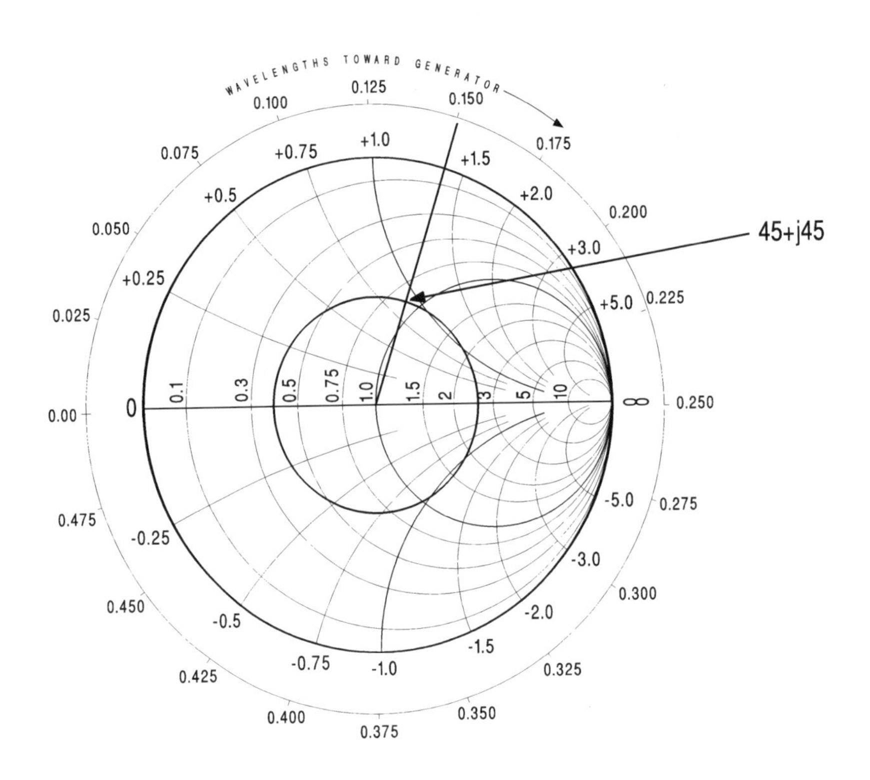

Examining the chart further, you can see that a quarter wavelength length of transmission line will convert any impedance into a corresponding impedance that is directly opposite it on the Smith Chart. It should be easy now to see why a quarter wavelength line converts a short to an open, and vice-versa!

A full revolution of the Smith Chart represents an electrical length of 0.5 wavelengths. This means that any impedance is "repeated" every half wavelength down a transmission line. Shorted or open lines that are not a multiple of a quarter wavelength represent either inductors or capacitors. The actual value can be determined by using the Smith Chart.

Now have a look at the complete Smith Chart that includes the horizontal graduations across the bottom that represent radially scaled parameters. The reflection coefficient Γ of a given load can also be read from a Smith Chart. Locate the normalized load impedance on the impedance chart, and draw a line from the prime centre through the point and out to the circumference. The angle of the reflection coefficient can be read directly from the circular scale, while the magnitude can be determined by measuring the distance from the centre to the point with a pair of dividers, then referring to the linear scale at the bottom of the chart. No math is required at all! Since S_{11} is the same as the input reflection coefficient, it can be easily plotted on the Smith Chart, and then the input impedance can be read off directly. Similarly S_{22}, the output reflection coefficient can be plotted.

The following Smith Chart illustrates the procedure. A bipolar transistor amplifier is being designed using a device with published S-parameters. S_{11} is given at three different frequencies: at 100 MHz it is 0.62∠-58, at 200 MHz it is 0.45∠-95, and at 500 MHz it is 0.35∠-141. Each of these values represents a magnitude and an angle in degrees. We will now plot these points directly on the Smith Chart:

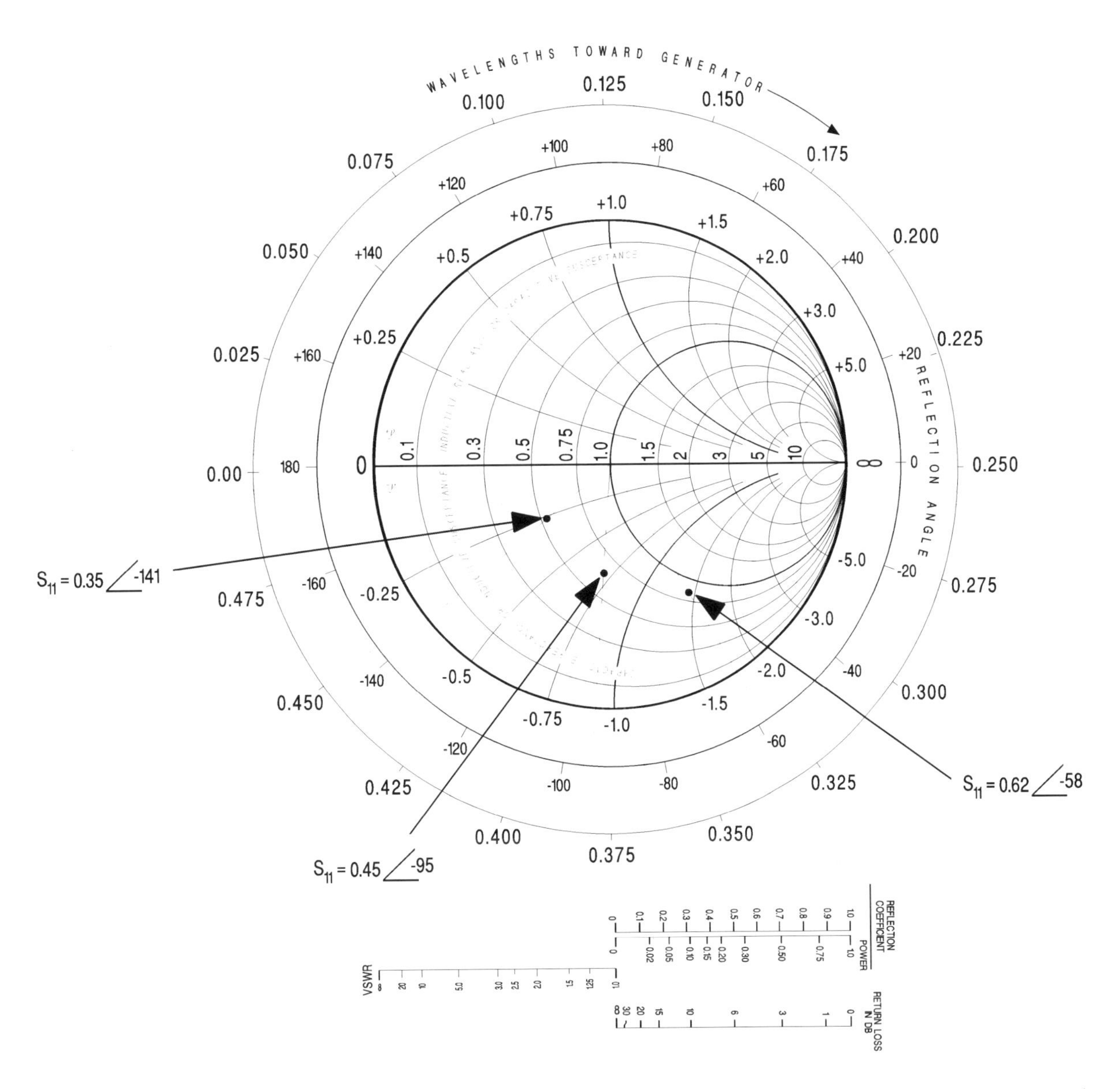

The three points above were placed by first drawing lines outward from the prime centre to the "Reflection Angle" scale at the angles of -58, -95, and -141 degrees. Then using the scale on the bottom of the chart labelled "Reflection Coefficient", distances from the centre were measured off representing the magnitude of S_{11}. The three points can now be converted to equivalent impedances by reading off the values from the other scales (don't forget to multiply by 50 ohms to normalize). The results are:

At 100 MHz, an S_{11} of $0.62\angle\text{-}58$ is equivalent to an input impedance of 41 - j72 ohms.

At 200 MHz, an S_{11} of $0.45\angle\text{-}95$ is equivalent to an input impedance of 31 - j35 ohms.

At 500 MHz, an S_{11} of $0.35\angle\text{-}141$ is equivalent to an input impedance of 27 - j13 ohms.

As we saw several sections ago, it is useful to think in terms of <u>admittances</u> rather than impedances when examining parallel circuits. The Smith Chart can quickly provide this conversion for us. Starting with the impedance, draw a line through the prime centre and across into the other half of the chart. Measure the distance from the centre to the plotted impedance, and mark a point on the extended line that is the same distance from the centre. This new point now represents the admittance coordinates!

An alternate approach to converting between impedance and admittance values is to overlay an admittance chart on top of an impedance chart. The admittance chart is just like an impedance chart, except that it is rotated 180 degrees. A single point can then be read off of either set of scales directly. When two superimposed charts are 180 degrees apart, the combined chart is called an **immittance chart**. The chart becomes a bigger jumble of lines and curves, and is a bit hard to follow unless the two different sets of coordinates are in different colours.

The effect of the addition of series components can easily be seen on the impedance chart. From the starting impedance, follow either a constant resistance or a constant reactance circle depending on whether the added component is reactive or resistive) to the new impedance. As an example, if the starting impedance is (30 - j20) ohms, this would be plotted on the normalized impedance chart as (0.6 - j0.4). Now assume that a series inductor of j40 ohms is to be added. The series inductor will add a normalized reactance of 0.8. From the starting point, follow the 0.6 resistive circle clockwise through a total of 0.8 reactive divisions, to end up at the point (0.6 + j0.4). This is shown below. Series capacitors work the same way, but rotating CCW. Series resistors move points toward the right.

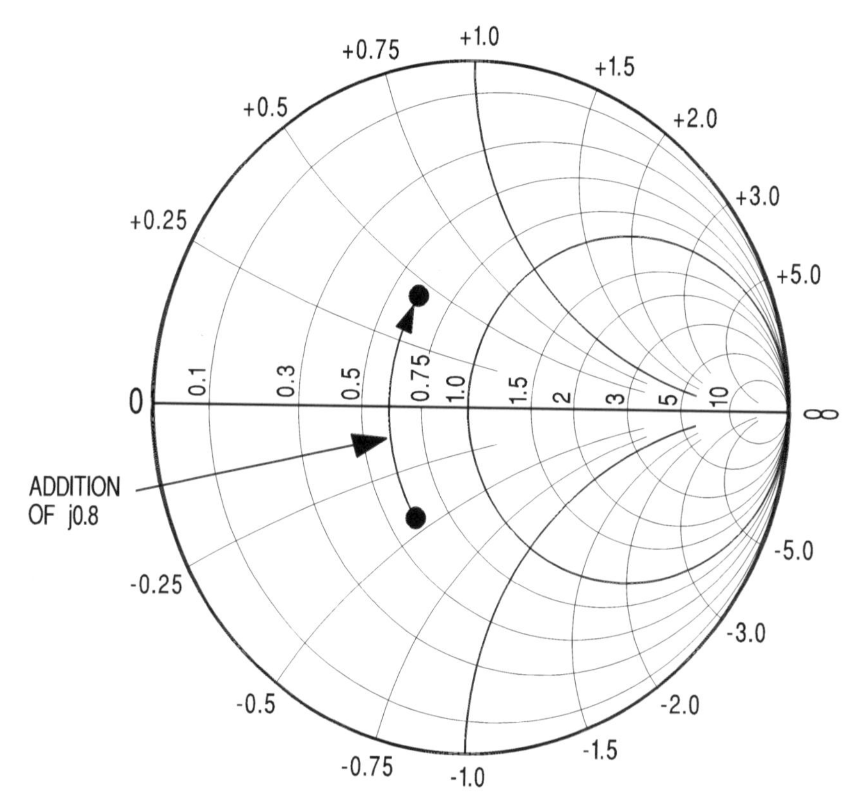

Any point on the Smith Chart has both resistance and reactance. Recall that the definition of component Q is simply X/R. It is therefore possible to determine the points having constant Q, and plot arcs on the chart connecting all points having the same Q.

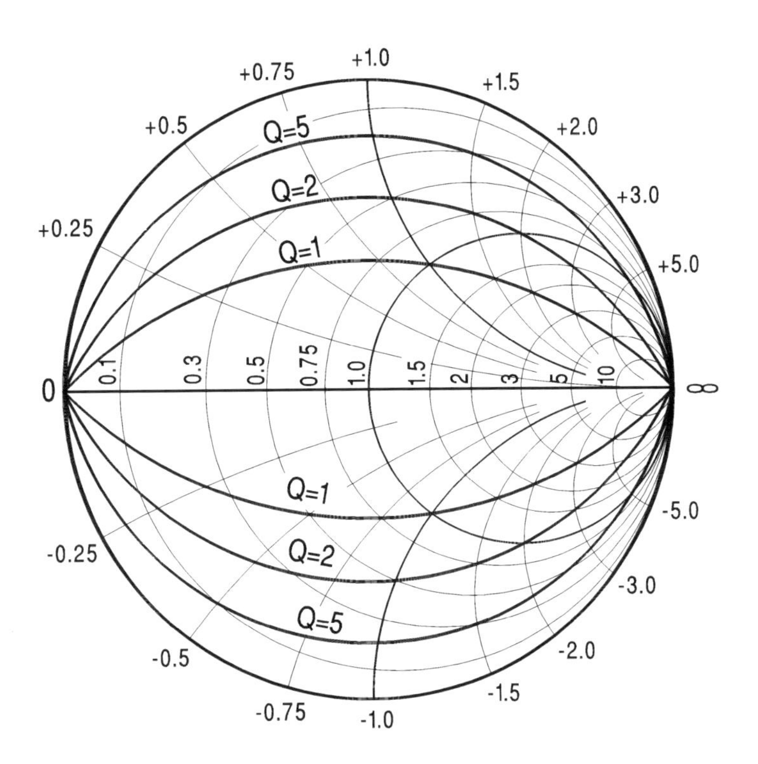

In the design of matching networks, it is sometimes desirable to keep the component Q's less than a specified value in order to maintain bandwidth and loss considerations. As long as a point is between the chart's centre and one of the constant Q arcs, the Q will be less than that of the constant Q arc.

MATCHING NETWORKS

Matching networks are simply circuits that transform one complex impedance into another. They can consist of combinations of inductors, capacitors, transmission lines, and transformers.

It is first of all important to understand why matching is needed at all. It has to do with optimum power transfer. A good analogy is the gearbox in a car. Assume that your engine can produce 200 horsepower at 4,000 rpm, and for simplicity, let's assume that it can produce exactly the same torque at all engine speeds. The engine's "red line" is also at a paltry 4,000 RPM.

When the clutch is first engaged from a standing start, the car has no forward motion, so the maximum torque at the wheels is required, and first gear is selected. Let's say that first gear has a ratio of 4:1. This means that the output of the gearbox will have one quarter of the input RPM, but 4 times the torque. The RPM is not important yet, because we are only starting to move. The high torque is ideal to accelerate the mass from a standing start. When first engaged, the engine's RPM is less than 4,000 and it is therefore producing less than 200 hp. As the car picks up momentum, the speed (both the car's and the engine's) picks up until when the engine RPM reaches 4,000 the engine is generating a full 200 hp, but the engine can't go any faster. We therefore had good acceleration as the engine sped up to 4,000 rpm, but now we can't use the torque that is available, because we have reached the RPM limit. For the moment, let's say that the actual car speed at this point is 30 mph. It only takes about 20 hp to maintain a car at a constant speed of 30 mph on a flat road, so we are only drawing one tenth of the available torque from the engine as it is screaming away at 4,000 rpm in order to maintain our vehicle speed. We now have a very **poor match** between the engine and the load (the car).

In order to continue building speed, we shift to second gear, which has a ratio of 2.5:1. The engine RPM now immediately drops to 2,500, and we have more torque available than is necessary to maintain 30 mph, so the car accelerates. When we have reached 48 mph, the engine has again reached 4,000 RPM, and we stop accelerating. We now shift into 3rd gear, which has a 1.5:1 ratio, and the process continues until we are traveling at 80 mph. At this speed, the car probably needs about 90 hp to propel it. We have 200 hp available, but we can't use it, because the poor match between the engine and the road has limited the engine RPM to 4,000 before we ran out of available power.

We therefore shift into top gear, which has a 1:1 ratio. We begin accelerating again, until at some vehicle speed the power required to maintain the car's velocity is just equal to that which can be produced by the engine at that RPM. In the ideal case, this will occur just at 4,000 rpm, and we can say that we have a perfectly matched condition between the engine and its load, and the vehicle speed is the highest that can be achieved with a 200 hp engine. If the load was not perfectly matched, either the engine's RPM limit would be reached before the car stopped accelerating, or the acceleration would diminish to zero before the RPM limit was reached.

In this simple example, the gearbox is acting as a **matching network** between the engine and its load. The gearbox is acting just like a transformer in these examples, where speed is equivalent to voltage, and torque is equivalent to current.

Now let's consider an electrical example. Assume that we have an AC voltage source which has an open circuit voltage of 100, and an internal impedance of 50 ohms. If we were to connect this to a load which was a pure resistance of 20 ohms, how much power would be delivered to the load? Treat this just as a simple voltage divider. The load voltage will be 20/70 times 100, or 28.6 volts. The power delivered to the load will be E^2/R, or 40.8 Watts. If the load resistance was 100 ohms, the load voltage would be 100/150 times 100, or 66.7 volts, and the power would be 44.4 Watts. If however the load was 50 ohms, the load voltage would be 50 volts, and the power delivered to the load would be 50 Watts. Here is an example of how maximum power transfer from a source to a load occurs when the load resistance is equal to the source resistance. In this simple example, the load is matched to the source when the load resistance is equal to the source resistance.

Now, let's complicate the example slightly. Assume that the load now consists of a 50 ohm resistor in series with a capacitor having a 50 ohm reactance at the operating frequency. The load impedance is therefore (50 - j50) ohms.

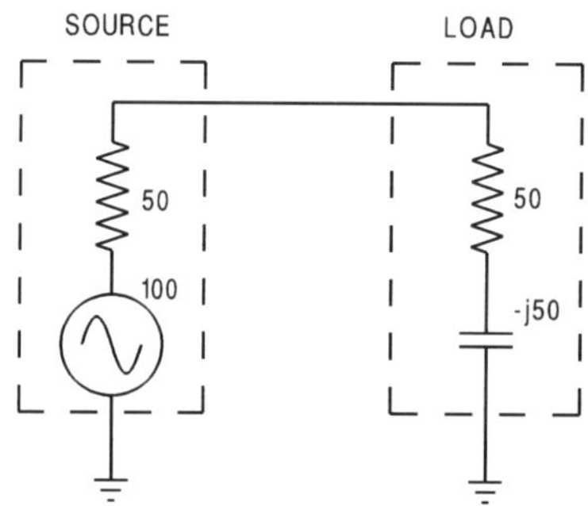

The ideal voltage source sees a total impedance of (100 - j50) ohms, and therefore the current will be: $I = \frac{100}{100 - j50}$ Amps. For the moment, we don't care about the phase angle, so we will deal only with the magnitude of the current:

$$|I| = \frac{100}{\sqrt{100^2 + 50^2}} = \frac{100}{111.8} = 0.894 \text{ Amps.}$$

The power delivered to the resistive part of the load is therefore $0.894^2 \times 50 = 40$ Watts. We would like to find a way of restoring the power in the resistive part of the load to 50 Watts.

We will insert a black box containing a **matching network** between the load and the source:

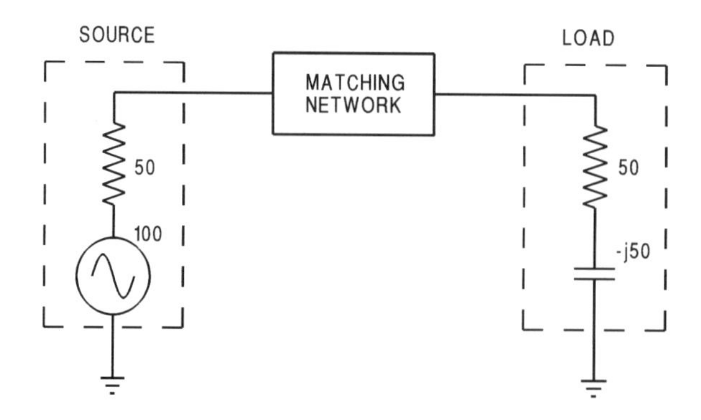

Since this is an easy example, we can immediately see what needs to be in the black box: a series **inductor** having a reactance of 50 ohms at the frequency of interest. Looking back at the source and matching network from the load, we will now see an effective source impedance of (50 + j50) ohms. The total series impedance that the voltage source will see is 50 +j50 + 50 - j50 = 100 ohms, and therefore the current will be 1 Amp. We have now restored the power in the resistive part of the load to 50 watts!

There is an important lesson to be learned here. The matching network transformed the 50 ohm source impedance to look like (50 + j50) ohms, which is the **complex conjugate** of the load impedance of (50 - j50) ohms. In matching complex loads, maximum power transfer will occur when the effective source impedance is the complex conjugate of the load impedance. This means that the resistive components are equal, and the imaginary component has the same value, but the opposite sign.

The trick in matching networks is therefore to make sure that all sources "see" a proper conjugate load impedance, and to make sure that all loads "think" they are being driven by a source with conjugate source impedance.

Unfortunately, very few RF components have 50 ohm resistive input and output impedances. In particular, the high frequency input and output impedance of transistors and FETs is usually quite low, and very reactive. Therefore, an appropriate matching network is required between most amplifier stages. Note that the network must provide a good match over the range of frequencies to be used.

If the impedance of a device is primarily "real" (meaning “resistive”), it can often easily matched by using a transformer. Using **bifilar** winding techniques, it is possible to make broadband small signal RF transformers that can cover the range from 1 MHz to well over 100 MHz. Low power devices are available at reasonable prices from companies like Mini-Circuits. The impedance transformation ratio is simply the square of the transformer's turns ratio. **Transmission line transformers** are exotic devices that can be used to match very high power transistor amplifiers at frequencies of up to about 50 MHz: these are often used in the output stages of commercial HF gear. Note that transformers are useful in matching source to load when the impedances are primarily resistive, but they cannot do the job alone if either the source or load has a significant reactive component.

For higher frequencies, especially when a small bandwidth (less than 20%) is required, a quarter wavelength of transmission line is a very effective way of matching between different impedance systems. The transmission line must have a Z_0 which is equal to the square root of the product of the source and load impedance. As an example, a 50 ohm transmitter can be readily interfaced to a 100 ohm load by using a quarter wavelength (don't forget the velocity factor) of transmission line having a Z_0 of about 71 ohms. At UHF and microwave frequencies, quarter wave transformers can be inexpensively fabricated using microstrip, and this is a common part of many matching schemes. It is easy to understand how this impedance transformation takes place by using a Smith Chart: plot source and load on a chart that has been normalized to Z_0 of the quarter wave line, then follow a constant VSWR around for 0.25 wavelengths.

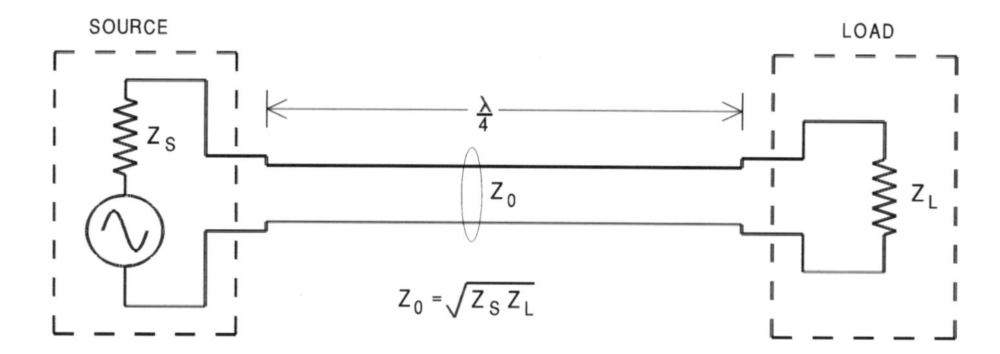

Networks using inductors and capacitors can also be used for matching between two differing impedances. We will start by considering impedances that are purely resistive. The three most common LC networks that are used in impedance matching are referred to as the L, T, and PI configurations. The names relate to the actual appearance of the component arrangement on a schematic.

The simplest LC matching network is the "L" network, and we will use it in an example to match from a 10 ohm source to a 50 ohm load.

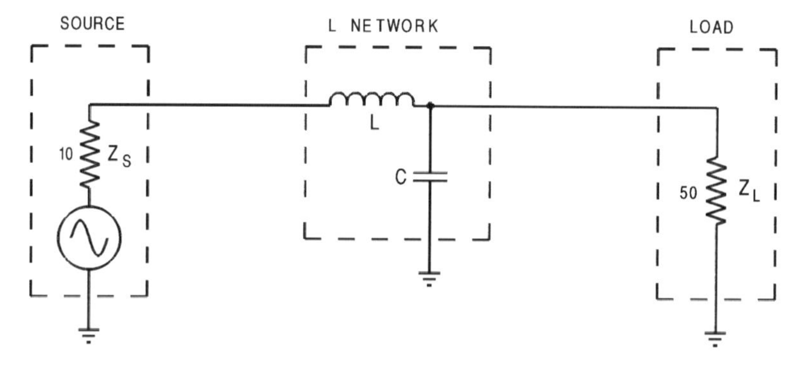

In order for this L network to perform a 5:1 impedance step-up, the reactance of the inductor must be 20 Ω at the frequency of interest, and the reactance of the capacitor must be 25 Ω. We will discuss where these values came from a little bit later. In order to understand how this circuit performs the impedance transformation, start out by looking toward the load from the junction of the inductor and the capacitor: you will see a 50 Ω resistor in parallel with a capacitive reactance of 25 Ω. This can be converted into an equivalent series circuit using the method from an earlier chapter, as shown below:

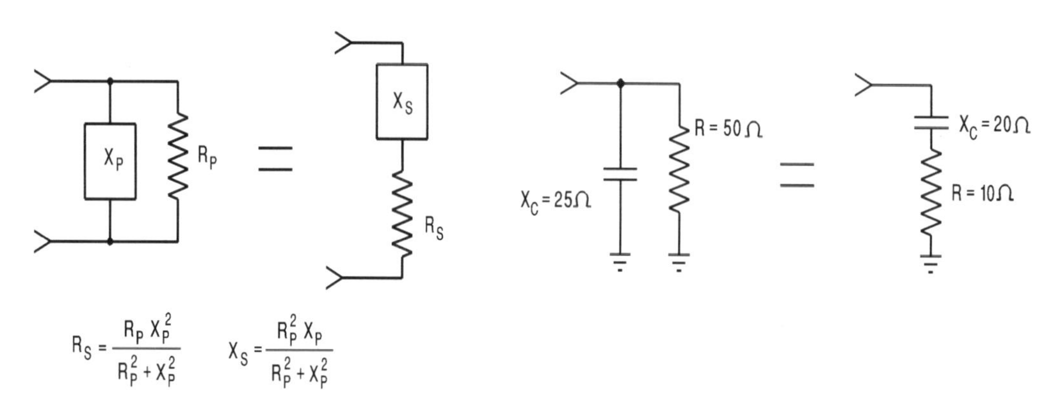

$$R_S = \frac{R_P X_P^2}{R_P^2 + X_P^2} \qquad X_S = \frac{R_P^2 X_P}{R_P^2 + X_P^2}$$

If the series inductor of the L network has a reactance of +j20 Ω, it will cancel the series capacitive reactance of -j20 Ω, and the source will only see what appears to be a 10 Ω resistor. Therefore, the source acts as though it is matched to 10 Ω, even though the actual load impedance is 50 ohms. The L network has therefore successfully transformed the 50 ohm load into a 10 ohm load for the source, and 100% power transfer can occur (ignoring stray losses). The actual values of the L and C in the network are chosen to give the correct reactance at the frequency of interest. Therefore, the match will only work correctly at a single frequency.

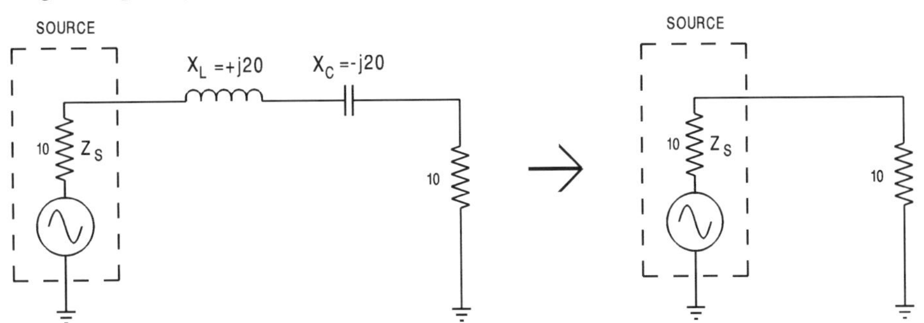

The actual values for an L matching network are determined by first of all determining the Q of the network: this is defined as $Q = \sqrt{\frac{R_L}{R_S} - 1}$. The capacitive reactance is then equal to R_L/Q, and the inductive reactance is equal to QR_S. The particular arrangement of this example L network constitutes a low-pass filter. If the positions of the L and C were reversed, we would have a high-pass filter. Either configuration can be used: the general design equations are as follows:

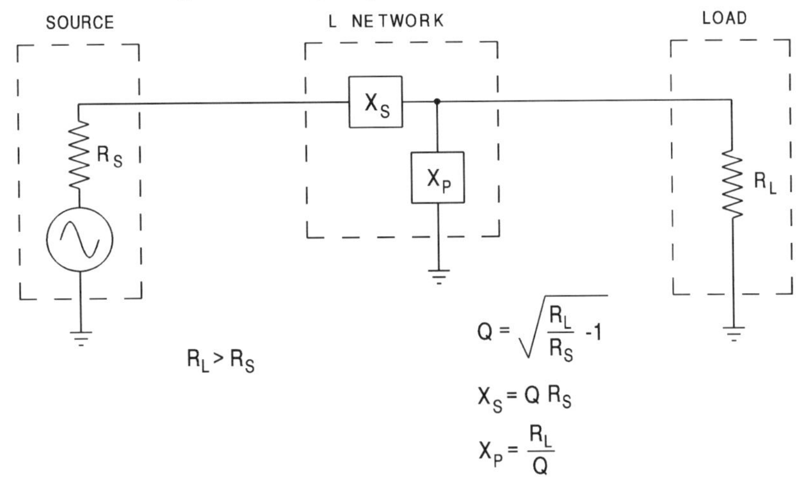

$$Q = \sqrt{\frac{R_L}{R_S} - 1}$$

$$X_S = Q\,R_S$$

$$X_P = \frac{R_L}{Q}$$

With this general purpose schematic of an L network, X_S and X_P can be either an inductor or a capacitor, but the two components must be different (in other words, there must be one L and one C).

If either the source or the load contained a reactive component, an L network could still be used to transform the real parts of the impedances, as long as the extra reactance is "absorbed" into the network (changing the value of X_S or X_P), or is somehow or other "tuned out" at the frequency of interest.

With this simple L network, the impedance can be "stepped up" in going from left to right. If it is required to "step down" an impedance, the network can simply be turned around. The designer doesn't have any control over the Q of the network - it is defined by the ratio of the impedances to be matched.

If a T network or Pi network are used, considerable more flexibility is possible. The output impedance can be either higher or lower than the input impedance, and the designer has freedom to select the network Q in order to provide the desired selectivity.

We will look primarily at the Pi matching network. This can best be understood by considering that it actually consists of two L networks that are connected "back to back". Both networks share the intermediate inductor:

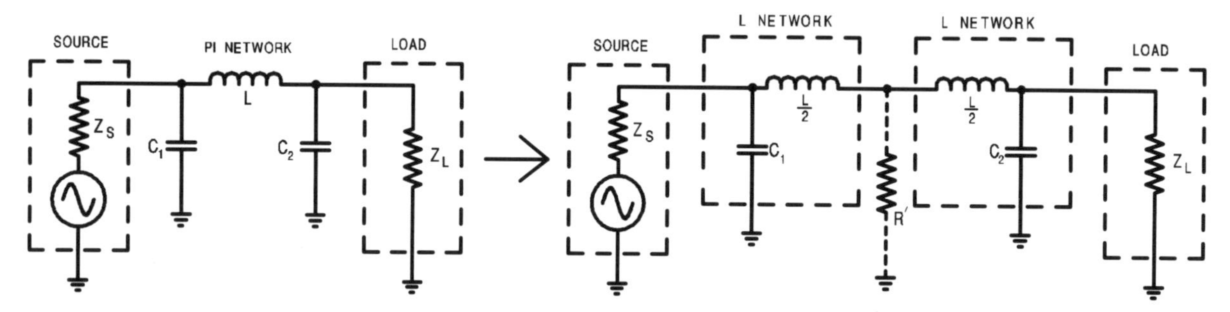

Here, the inductor of the Pi network has been "split in half", forming two L networks. R′ is a "virtual resistance" that can be thought of as existing between the two networks. Starting from the source, the first L network steps down the impedance to R′, and then the second L network steps the impedance down up to the final output impedance. The assumed value of R′ can be chosen in order to give the desired value of Q. Again, the actual components are shown in a low-pass configuration - they can also be rearranged as a high-pass structure. The schematic for a generalized Pi network is:

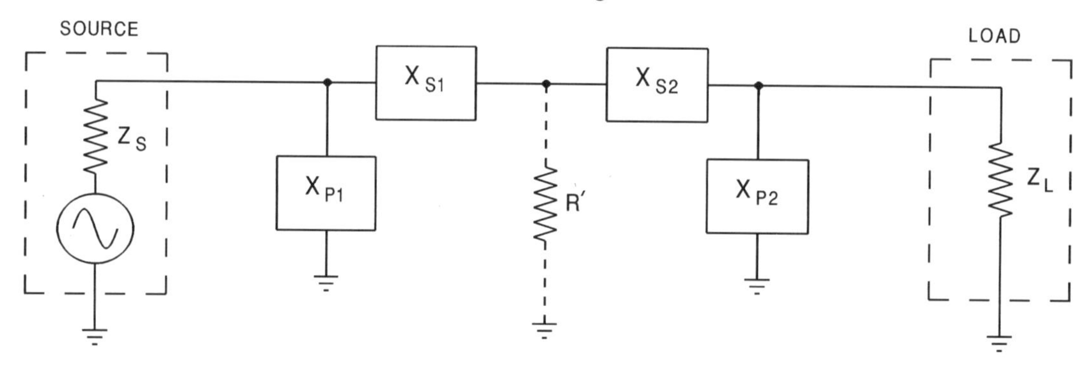

The design procedure using this generalized schematic is as follows:

1. Choose a desired value of network Q.
2. Define R_H as the highest value of Z_S or Z_L.
3. Determine the value of the virtual resistor as $R' = \frac{R_H}{Q^2 + 1}$

4. Use the L network design equations to determine the values of X_{S1} and X_{P1} that will match Z_S to R'.
5. Use the L network design equations to determine the values of X_{S2} and X_{P2} that will match R' to Z_L.
6. Add the values of X_{S1} and X_{S2} to determine the total series reactance required.
7. Compute component values to give the required reactances at the frequency of interest.
8. If the load or source impedance contains a reactive component, either absorb its value into the network, or add an additional component to tune it out.

Smith Charts are very useful in determining the necessary reactance of the matching components. There are several different approaches that can be used, but we will illustrate one straight-forward technique by using a simple example.

Let's assume that we have a device whose input impedance is (20 + j100) ohms. We wish to match this to a 50 ohm source. We start by normalizing the impedance: divide by 50 to get (0.4 + j2.0), then plot the normalized impedance on an **immittance chart**. Note that if we look at the scale representing admittance, we can immediately read off the equivalent admittance data as (0.096 + j0.48). We wish to convert the point on the Smith Chart to a point right at the centre, which represents 50 ohms. Because an immitance chart is very difficult to read if it is printed in a single colour (as required for this book), note that the figure shown below only duplicates a portion of the chart.

Refer to chart on the next page.

Starting from the plotted point, we will use the admittance overlay to follow a constant conductance circle in a clockwise direction, simulating the effect of the addition of shunt capacitance. We will continue to move in this direction, following the circle representing a conductance of 0.096 until we intersect the underlying circle representing a resistance of 1.0. This occurs after the addition of a normalized susceptance of (0.48 + 0.28) = 0.76. We have therefore simulated the addition of a shunt capacitance at the load with a value who's susceptance at the frequency of interest is 0.76 x 50 = 38 mS.

We now transfer to the impedance scales, and note that this point represents an impedance of (1.0 - j3.2). Well, we have got the resistive component to where we want it, now all we have to do is take care of the imaginary component. This can be handled by a series inductor. If an inductor is placed in series whose reactance at the frequency of interest is 3.2 x 50 = 160 ohms, we will have completed the match back to 50 ohms.

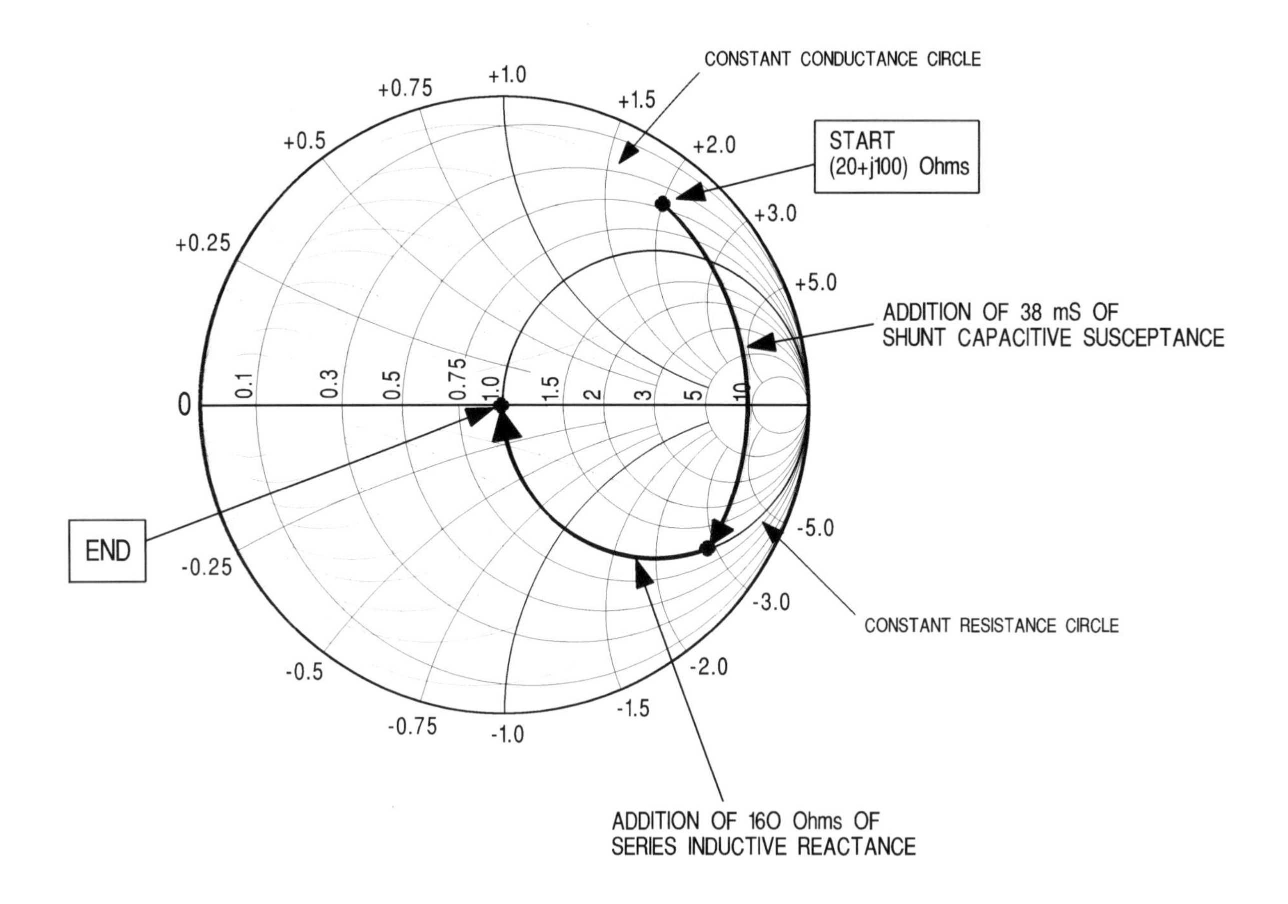

The physical circuit will therefore look like our familiar L network. In this case, the load has a reactive component, but the matching network has absorbed the load reactance. The shunt capacitance has a susceptance of 38 mS, which can be inverted to get a reactance of 26 Ohms.

Continuing with our example, the final circuit therefore has the following reactance values:

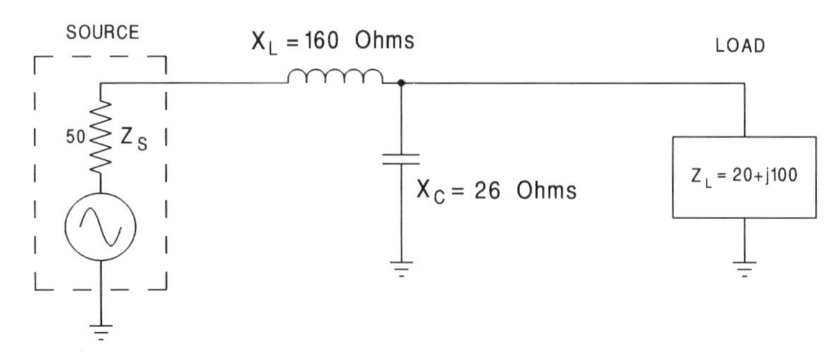

Actual component values are a function of the frequency. Either the reactance formulas or the reactance chart can be used to determine the inductance and capacitance.

Depending on where the starting point is on the Smith Chart, different combinations of series and parallel L and C can be used to convert any point back to the prime centre. This is shown below:

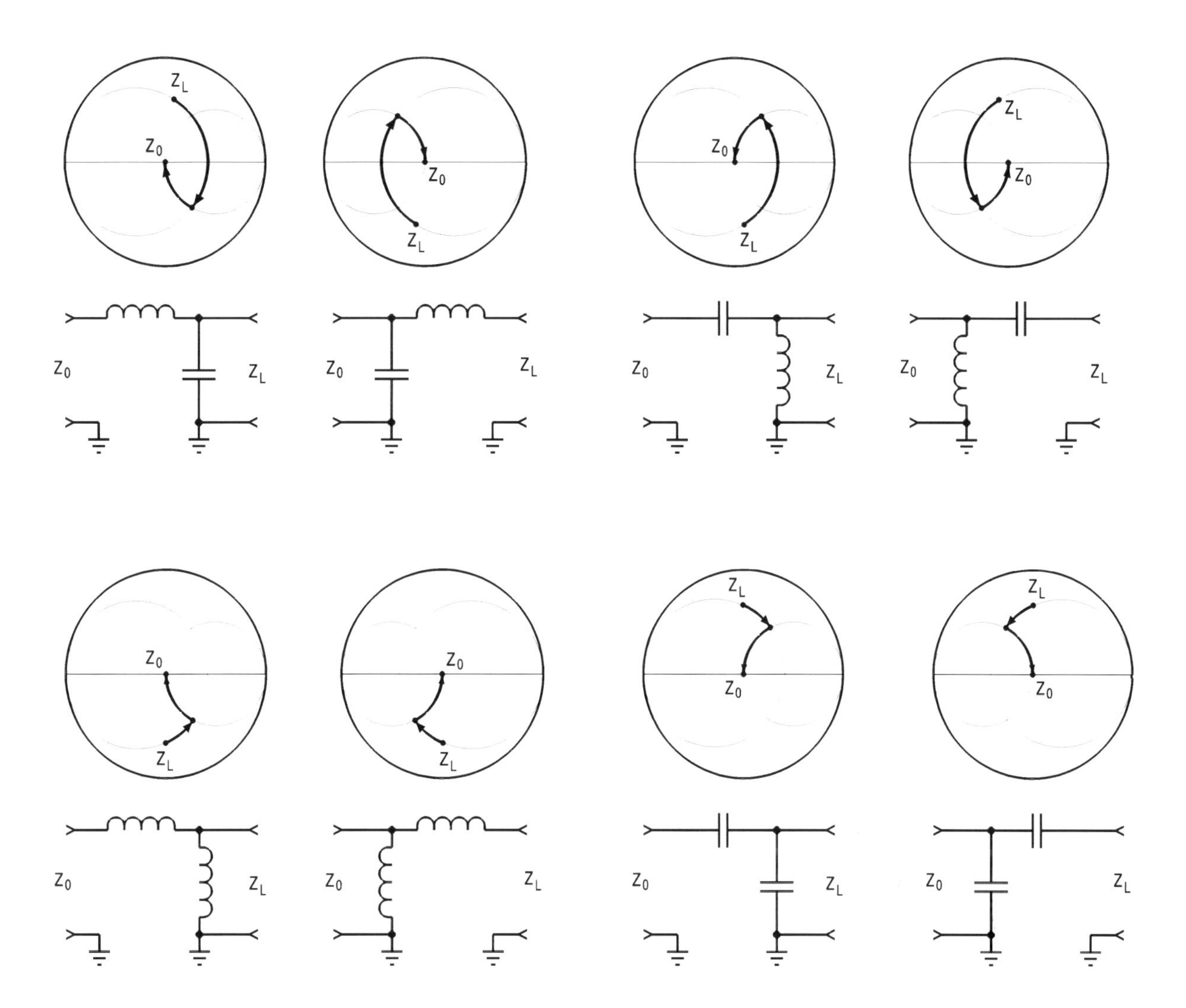

At UHF and microwave frequencies, the Q of discrete components is often not sufficient to allow these techniques to yield low loss networks. You can overlay arcs of constant Q on the immitance chart and examine the Q of intermediate points in the matching network. If Q is much over 4, losses can be appreciable at frequencies much over 1 GHz. However, at these high frequencies, sections of transmission line are easy to incorporate into the network, as they can be made part of the circuit board. Having the freedom to use L, C, and lengths of line gives great flexibility: For any impedance to be matched, there are often several effective circuits that can be used.

This has been just a glimpse at the use of Smith Charts for designing matching networks. The interested student is urged to review the book "RF Circuit Design" by Chris Bowick (WB4UHY) for much more detail. Several software programs (some free) are available on the World Wide Web that can aid in the design process. The true aficionado will also want to read Phillip Smith's book, entitled "Electronic Applications Of The Smith Chart".

MIXERS, MULTIPLIERS, AND MODULATORS

As part of basic radio education, everyone learns that a mixer can be used to produce sum and difference frequencies from two input frequencies. With an appropriate output filter, the desired output can be selected.

A single balanced mixer or a doubly balanced mixer also produce sum and difference output frequencies, but they also suppress the output of either the local oscillator signal or the RF input signal, or both.

We will now try to understand how the mixing operation actually works. If we have two RF signals at frequencies f_1 and f_2, how are the sum and difference frequencies generated? If these two signals are simultaneously fed into the input of a linear amplifier stage, the output will have f_1 and f_2, but will not have any traces of the sum and difference frequencies if the amplifier is perfectly distortionless.

Let us now consider an analog multiplier circuit. At any point in time, the output voltage is equal to **the product** of the two instantaneous input voltages. There are several integrated circuits (such as the MC1495 or the AD734) which can perform this function at medium frequencies. If we hook up a power supply to one of the inputs, and put a 1 volt RF signal into the other input, the output will be an RF signal of the same frequency as the input, but with an output voltage which is proportional to the DC voltage on the first input. If a 1 volt DC input is put into port 2, the RF output will be a replica of the RF input. If a -1 volt input is put into port 2, the RF output will be just like the RF input, except that it will be 180 degrees out of phase. The multiplier can thus be used as a controllable amplifier or attenuator. Now consider the situation where we put one RF signal into the first input of the multiplier, and another RF signal with a different frequency into the second input. The output will therefore be $\cos(\omega_1 t) \cdot \cos(\omega_2 t)$. This looks like an ungainly expression: how do you multiply trig functions? Fortunately there is a trigonometric identity that can be used here:

$$\cos(\omega_1 t) \cdot \cos(\omega_2 t) = \frac{1}{2}\cos[(\omega_1 + \omega_2)]t + \frac{1}{2}\cos[(\omega_1 - \omega_2)]t$$

This is a very interesting result! This says that the output consists **only** of the sum and difference frequencies: it does not have outputs at either of the two input frequencies.

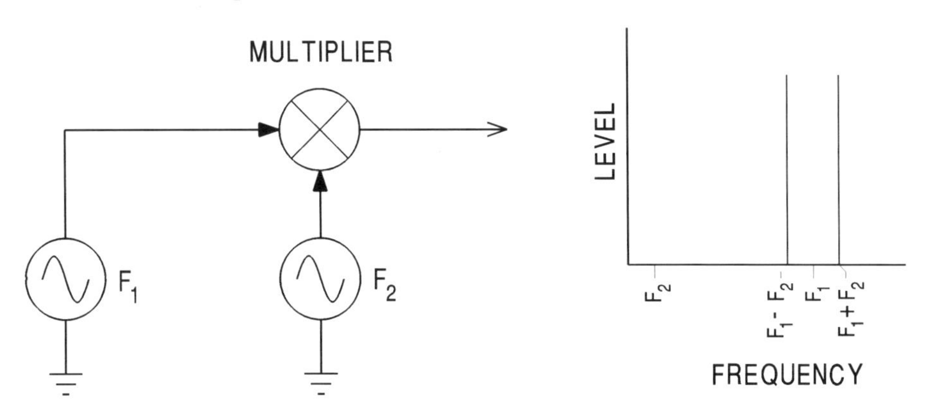

If one of the inputs is RF, and the other is AF from a microphone and speech amplifier, we have just created a double-sideband suppressed carrier output. If an appropriate sideband filter (probably either a very sharp crystal or mechanical filter) is added, we can now generate an SSB signal. We have just learned that the multiplication function is very useful for mixers or modulators.

In order to generate an AM signal, we should add a DC component to the audio input to the multiplier. If the DC voltage is equal to one half of the maximum peak-to-peak audio signal, an AM signal with 100% modulation will be generated.

Multipliers are nice, but there are even simpler ways of performing very basic mixing operations. It turns out that any non-linearity can produce sum and difference frequencies. Consider for the moment applying two RF signals to a component whose output is equal to the square of the input voltage. Again, there is a trigonometric identity that can be effectively used here:

$$\left(\cos(\omega_1 t)+\cos(\omega_2 t)\right)^2 = 1+\frac{1}{2}\cos(2\omega_1 t)+\frac{1}{2}\cos(2\omega_2 t)+\cos[(\omega_1+\omega_2)t]+\cos[(\omega_1-\omega_2)t]$$

Examining this equation, it can be seen that the output consists of a DC component, plus the second harmonics of the two input frequencies, plus the sum and the difference frequencies. If the component's response did not exactly follow a square law relationship, the output would also contain components at the two input frequencies. Now recall that the current through a diode is roughly an exponential dependent on the applied voltage. Therefore, if two RF signal voltages are injected into a diode, it is possible to extract sum and difference frequency components from the diode's current. The output will also contain the two input signals, as well as their harmonics and some other mixing products - a filter is needed after the mixer to select the desired signal.

FET switches or CMOS transmission gates can also be used as mixers. Typically the lower frequency is in analog form, and the higher frequency is a digital square wave. Consider the following circuit:

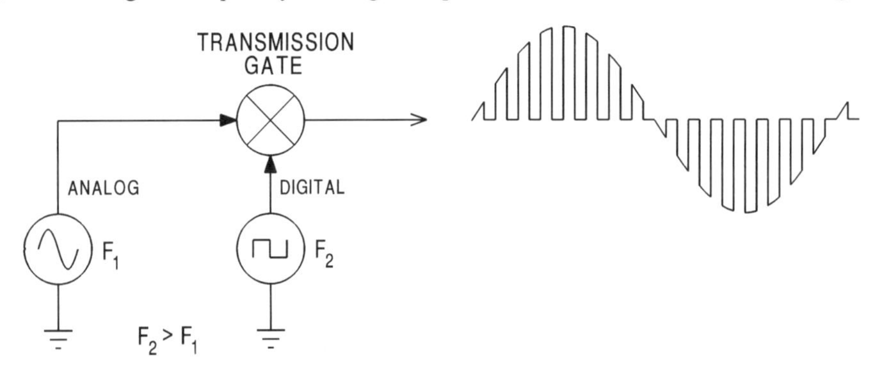

If the output of the transmission gate is filtered to remove all harmonics of F_2, the output spectrum of this quasi-digital mixer will be just like that from a conventional analog balanced mixer - this is because of the multiplication action. The output will only contain signals at the sum and difference frequencies.

Apart from understanding how mixers work, remember that any distortion in an amplifier stage will cause the input frequencies to mix and generate sum and difference frequencies. Distortion will also cause the generation of harmonics, as we learned earlier. Therefore the output spectrum of a non-linear amplifier that is being fed two differing frequencies will contain not only the original frequencies, but also harmonics of them, as well as sum and difference frequencies of the fundamentals and their harmonics - there are a lot of frequencies present at the output. Keep this in mind when we discuss "Intermod" later in this book!

DIGITAL LOGIC

General purpose digital logic does not normally involve the square root of minus one, standing waves, Smith Charts, or complicated trigonometric identities. It is involved with only two signal levels: "1", and "0". When a voltage is supposed to change from a "0" to a "1" or vice versa, the object is to cause the transition to occur as rapidly as possible: linearity is not an issue. The definition of what voltage constitutes a “1” and what constitutes a “0” is a function of the logic family and the power supply voltage.

The two most basic logical functions are the "**or**" and the "**and**". Considering a two input device, the "or" gate produces a "true" output if either input is true, while the "and" gate requires that both inputs be "true". The term "true" usually refers to a signal being at a "1" state, or the higher of the two logic voltages, while the "false" state is represented by "0", and is the lower of the two voltages.

The two fundamental types of logic gates are the "and" and the "or" function. It is easy to think of "and" and "or" gates in terms of switches and relays. Two SPST switches wired in parallel between a voltage source and a load represent the "or" function: the load is active if either switch is closed. Two SPST switches wired in series between a voltage source and a load represent the "and" function: both switches must be closed to activate the load.

The "and" function is written as the operator •. Therefore, if it is desired to describe the function "A and B", it is written as A•B, or more simply "AB". The "or" operator is written as "+". Therefore the function "A or B" is written as "A+B". If the inverse function is intended, it is written as "NOT": a function indicating "NOT A" is abbreviated as $\overline{A}$. Therefore the example $\overline{A}+B$ means "B or NOT A", to distinguish it from the expression $\overline{A+B}$, which means "NOT A or B".

A special function called the "exclusive or" function is written as " $\oplus$ ". The output of an exclusive or gate is true if only one of the inputs is true. If both of the inputs are true, or both are false, the output is not true.

A very simple function is the "inverter". The output state is always the opposite of the input state.

The schematic symbols for these basic logic devices together with their logic equations is shown below:

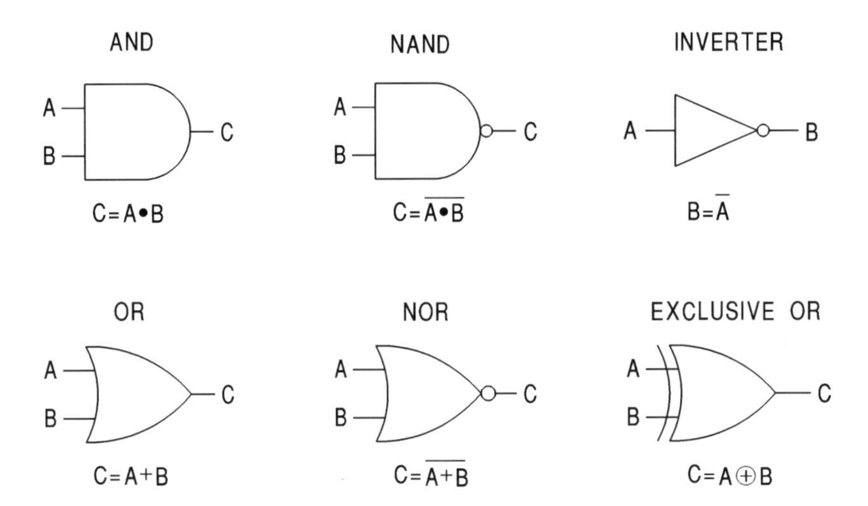

A simple "or" gate can be made with two diodes and a resistor:

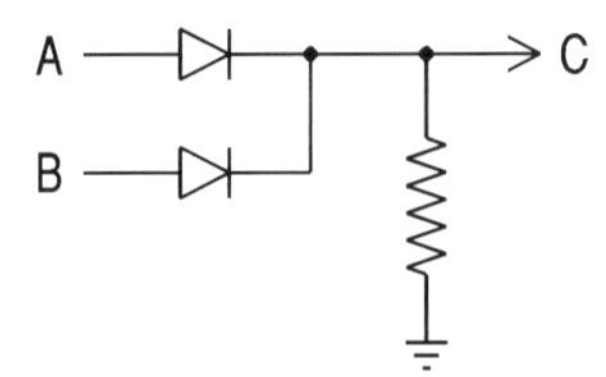

Note that the output is high if either of the inputs is high. Obviously, more diodes can be used to create more "or" type inputs. This basic concept was used in a series of logic devices offered in the mid-60's which was called "DTL". The diodes fed into the base of a transistor, and the output was taken from the collector of the transistor. Now, consider this identical circuit, but instead think of the two inputs in terms of their inverse. This means that if both $\overline{A}$ and $\overline{B}$ are in their "0" or "false" state, they are at a high voltage. The output C will clearly be high unless both the inputs are "true", which means they are both at a low voltage. In other words, the same circuit will act as an "and" gate when considering negated inputs. This is an important observation, and is fundamental to **DeMorgan's Theorem**.

Much of digital logic is designed to work on 5 volt power supplies, and has logic levels of somewhere below 1 volt representing a "0", and somewhere over +3.6 volts representing a "1". The DTL family was actually preceded by RTL logic ("resistor-transistor-logic"), which did not use the isolation diodes. This was one of the first commercial logic families in the early 1960's.

In the late 60's, TTL logic (transistor-transistor-logic) was introduced. It offered higher speeds and many other advantages over DTL. TTL flip-flops could operate at speeds up to about 20 MHz. The TTL logic family was widely adopted: Texas Instruments was one of the primary suppliers. Most part numbers begin with the numeric prefix "74" (for commercial grade) or "54" (for military grade). The basic "and" gate is the 7400, while an inverter is a 7404. TTL is fairly power-hungry, so a lower power variant called the "74L" family was made available: it operated at roughly half the speed of standard parts. For speed freaks, the "74H" family was introduced: it increased speeds by about 30%, but used even more power. In the early 70's, a family of "74S" parts was introduced: this used a schottky diode in the output structure to reduce saturation problems, and speeds were increased without any increase in power. The next step was the "74LS" families, which offered high speed and low power. Using "74S" parts, it is possible to work with clock speeds of up to 50 MHz.

Motorola pioneered the commercialization of ECL (emitter-coupled-logic). This is a very high speed logic family that keeps the switching transistors out of saturation. The difference between a "1" and a "0" is approximately 1 volt. The basic ECL structure is a differential amplifier. ECL is normally designed to operate from a minus five volt power supply, but it is not difficult to use it at the normal logic supply voltages of 0 and +5 volts. Most devices have complementary outputs (one output is the opposite of the other). This is very high speed logic: clock speeds of over 125 MHz are common. The first three Motorola logic families were called MECLI, MECLII, and MECLIII: each family was faster (and more power-hungry) than the previous. In the early 70's, a new family called MECL 10,000 was introduced which offered reasonably fast speeds with medium power dissipation. Other manufacturers also produce ECL. Some of the fastest ECL came from Plessey and a few Japanese vendors: clock speeds of over 3 GHz are available. A level translator is necessary to interface ECL with TTL devices. Assuming that both logic families are operating from the same +5 volt power supply, the following circuit works quite well:

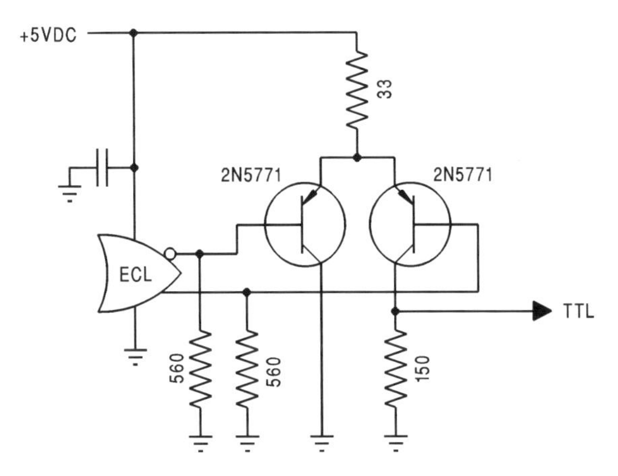

High frequency design techniques must be used with ECL. Every chip needs it own low-impedance power supply bypass, and all logic runs must be made using microstrip and proper termination resistors. Some parts dissipate considerable heat, and cooling is a consideration: moving air or clip-on heat sinks are sometimes necessary.

Various vendors developed logic components based on P-MOS and N-MOS FETs in the late 60's. RCA pioneered the use of both P-type and N-type devices in the same circuit, using a complementary output structure: they called this logic family "CMOS". Apart from reasonable clock speeds vs the previous MOS generations, the big advantage was lower power dissipation. Indeed, the quiescent (ie static) power consumption is virtually negligible. RCA developed an extensive family called the "4000 series": a basic "and" gate is a 4011. Many vendors now offer 4000 series parts. This logic family can operate on power supplies between +5 and +15 volts: clock speeds are higher at the higher supply voltages. Several vendors now offer a family of "74C" parts: they use the same part numbering and pin connections as TTL parts.

It is possible to connect the output of a CMOS device directly into the input of an LSTTL gate if they are operating at the same supply voltage. In order to reliably interface the output of a TTL device into the input of a CMOS gate, a "pull-up" resistor is required to ensure a solid "1" level.

Many very complicated functions are available in CMOS. Most microprocessor families and accessory devices are built using this logic structure. New logic families are becoming available which are designed to operate from a power supply of +3.3 volts. The newer microprocessors are CMOS devices that operate on +3.3 volts (or in some cases lower).

Some TTL and CMOS devices actually have three logic states: "1", "0", and "Hi-Z". In the Hi-Z mode, the output is virtually open-circuited: this is useful for putting multiple memory devices onto parallel data buses. This is sometimes called "tri-state logic".

There are many formalized methods of taking the truth table requirements for a logical function and allowing it to be reduced to the simplest possible assemblage of interconnected gates. A few easy rules to understand will help in simplifying logic diagrams:

$$A + A = A$$
$$A \bullet A = A$$

$$A + 1 = 1$$
$$A \bullet 0 = 0$$

$$A + AB = A$$
$$A \bullet (A + B) = A$$

$$A + (B + C) = (A + B) + C$$
$$A \bullet (BC) = (AB) \bullet C$$

$$\overline{\overline{A}} = A$$
$$\overline{0} = 1$$
$$\overline{1} = 0$$

$$\overline{AB} = \overline{A} + \overline{B}$$
$$\overline{A + B} = \overline{A}\,\overline{B}$$

This is **DeMorgan's Theorem**

Note that the last two equations indicate that a NOR gate can also be used as a NAND gate simply by changing the polarity of the inputs!

Exclusive OR gates are available as a complete IC (the 7485 contains 4 of them), or one can be made from a quad NAND gate such as a 7400. This is possible because of judicious application of the above rules of **Boolean algebra**. The "exclusive nor" function can be written as:

$\overline{C} = \overline{A}\,\overline{B} + AB = (AB + \overline{A})(AB + \overline{B})$. The actual circuit looks like this:

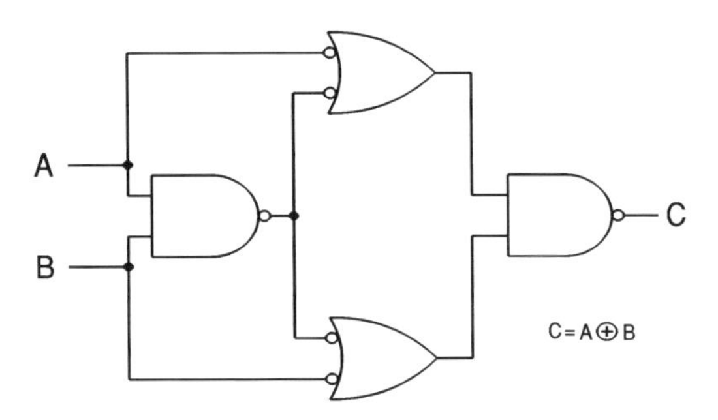

In the above circuit, each of the four gates is a two-input Nand gate. Two of the gates have been drawn as nor gates because their inputs are defined as using inverted inputs. This is just another application of DeMorgan's theorem.

Flip-flops are another major logic building block. They have two states: "1", or "0". A flip flop will stay in its state unless it is caused to change state by an external signal. Flip-flops are therefore commonly used in semiconductor memories. A simple Set-Reset flip flop can be made from two cross-connected gates:

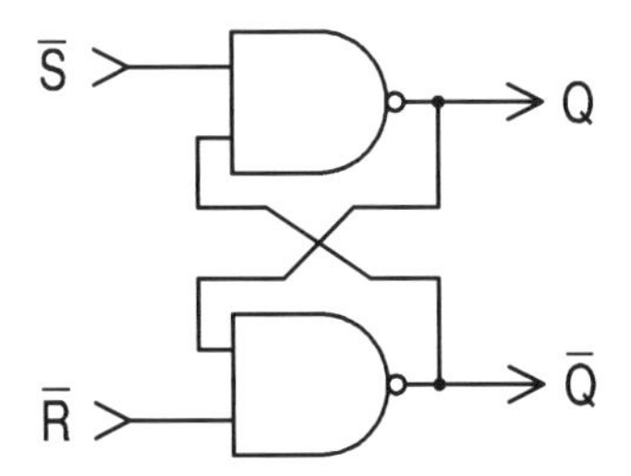

If both inputs are high, no change will occur. If either one of the inputs goes low (even momentarily), the state of the flip flop will be forced to the corresponding value. This particular implementation (using cross-coupled gates) is useful for building push-button de-bounce circuitry.

There are three basic flip-flop types that are available in logic IC's: the "T-type", the "D-type", and the "J-K type". Most are manufactured using a master-slave structure, which basically means that the output will not change until a full cycle of the input clock has occurred, as opposed to just a single edge transition. The schematic symbol for the three types look like this:

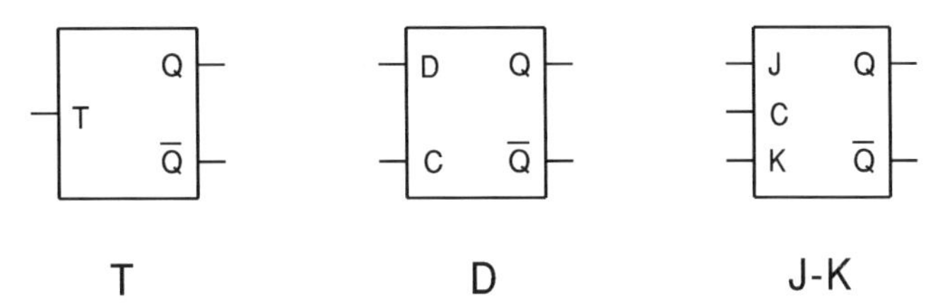

A T-type flip flop changes to the opposite of its current state after every cycle of the clock. A D-type transfers the state of the D input pin to the output after a cycle of the clock. The 7474 is typical of a D-type flip flop.

A J-K type operates in several different ways. If both J and K inputs are held high, the device operates just like a T-type flip-flop. If both J and K are held low, the device will not change state after a clock cycle. If either the J or K input is held high while the opposite input is held low, the state will change to match the inputs (sort of like a D-type). The 7473 typifies a J-K flip-flop.

Most flip-flop IC's also have direct Set and Reset pins that can be used to force the state independent of the clock.

If a continuous square wave is fed into the input of a T-type flip flop, the output frequency will be one half of the input. By interconnecting flip flops and gates, it is possible to build circuits that will divide the input frequency by any integer number.

The material in this chapter up to this point was written in the mid-90's. The theory is completely relevant today, and many of the logic families described are readily available. However, 20 years later we find that more and more logical functions are being performed with small, low cost microprocessors. Where very high performance digital functions are required, Field Programmable Gate Arrays (FPGAs) are often the device of choice.

FREQUENCY MULTIPLIERS AND DIVIDERS

Frequency multipliers are devices whose output frequency is an integral multiple of the input frequency. In other words, the output is a harmonic of the input frequency. The easiest way to generate harmonics is to cause distortion on the input signal. The appropriate harmonic can be selected by a band-pass filter and then amplified. As we saw during the analysis of mixers, a diode is useful for generating second harmonics. It is also useful for higher harmonic numbers. The energy from any harmonic-generating device falls off at higher harmonic numbers.

An amplifier can generate lots of harmonic energy if it is biased so that it is cut-off over an appreciable portion of the input cycle. Push-Pull multipliers are good at generating odd harmonics, but have little even harmonic energy. A "Push-Push" configuration generates plenty of even harmonics.

Varactors (voltage variable capacitors) can also be used to generate harmonics. Varactor multipliers are often used to multiply medium power (less than 25 Watts) VHF signals up to the UHF range with decent efficiency.

If the input frequency is converted into a square wave with small rise and fall times, appreciable energy can be found at odd harmonics.

A T-type flip-flop can be used to divide a frequency by 2. Obviously, multiple flip-flops can be cascaded to generate a division ratio of 2^N. When flip-flops are cascaded, by examining the actual state of each individual flip-flop, you will observe that a binary sequence is being followed: if the parallel outputs are used to define a digital word, you have created a **counter**. Binary frequency counters using exotic ECL are available to divide frequencies of over 2 GHz.

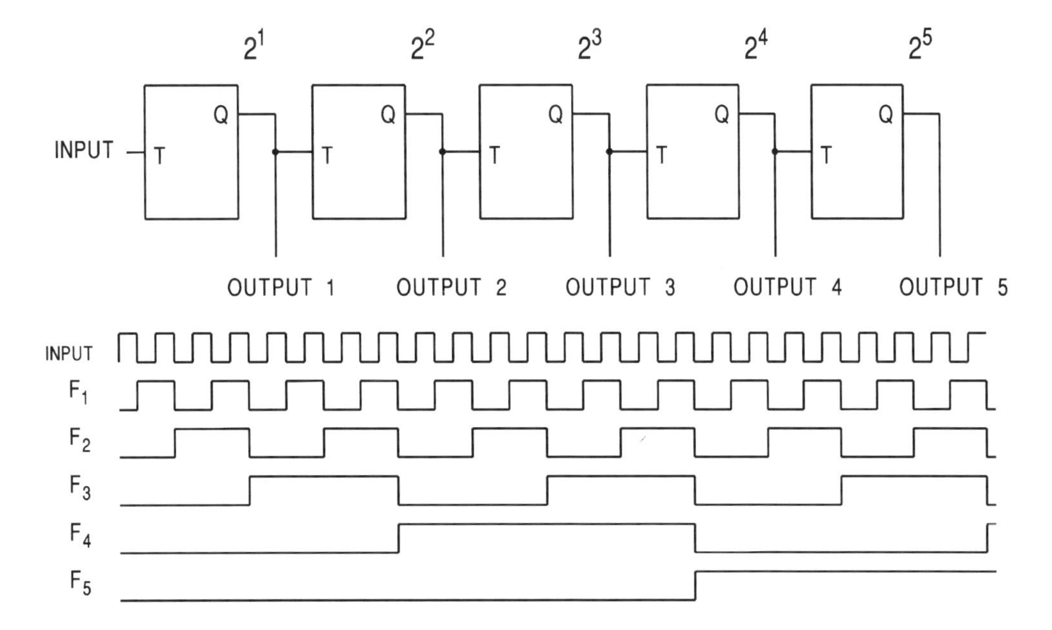

To create counters that divide by ratios which are not binary, external gates can be used to decode specific combinations of states and cause a reset to occur. A divide by-10 counter such as the 7490 consists of a divide-by-2 stage followed by three flip-flops and some gates to create a divide-by-5 circuit. By cascading 7490's (or equivalent parts), you can create division ratios of 10^N.

Dual-modulus counters are counters that either divide by the preset ratio or by one more than the ratio, depending on a control input. If the control input itself is driven by a divided-down clock, the net effect is to produce an operation similar to that which would be provided by a divider that could divide by non-integer numbers.

Counters are available that can be preset to a specific state. This can be used to shorten the counting sequence. Counters are also available where the division ratio is defined by external parallel digital inputs. These are of great use in frequency synthesizers.

A variety of high speed CMOS counters are available that are completely programmable, and are often integrated with other components that will be required to implement a synthesizer.

THE PHASE-LOCKED LOOP

As we have just seen, it is possible to use multipliers or counters to increase or decrease a frequency, but only in integral numbers. This is somewhat limiting. It is often desired to use a high stability reference frequency to generate a another frequency that is not harmonically related to it but has the same stability characteristics. A circuit known as a phase-locked loop (or "PLL" for short) can be used to accomplish this. A generalized block diagram of a PLL is as follows:

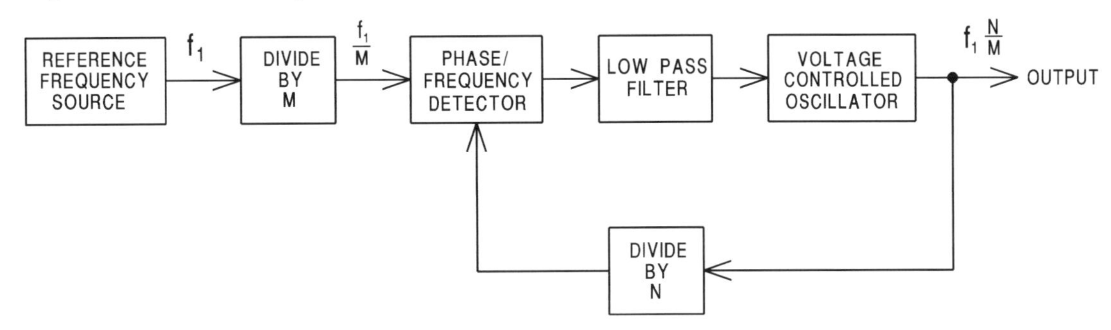

The frequency/phase detector will generate a DC output signal that is high if the reference frequency is higher than the output of the divider, or is low if the reference frequency is lower than the divided voltage-controlled-oscillator (VCO) frequency. This component has many similarities with a balanced mixer which has a DC output.

This is a feedback system. The loop will be "locked" when the input frequencies and phase of the reference and the divider output are identical. If they are not identical, the DC voltage will move in such a direction to force lock to occur. The output frequency is equal to the input reference frequency times the ratio N/M, where both N and M are integers. The fractional stability and accuracy of the output frequency will be identical to that of the reference. The output frequency can be changed by altering either N or M (or both).

A typical challenge might be to design a circuit that starts with a 1 MHz input from a stable crystal oscillator, then can be used to generate frequencies from 20 to 30 MHz, with 10 KHz channel spacing (1,000 channels). This clearly is impractical to do using just dividers and counters, so we will use a PLL configured as follows:

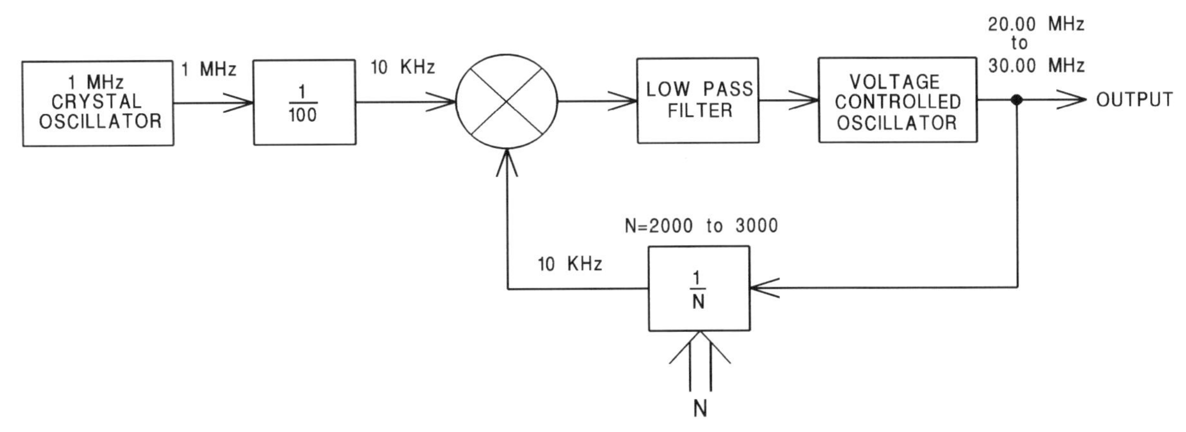

Note that the 1 MHz precise frequency is divided down to create a reference frequency that is equal to the required channel spacing of 10 KHz. The VCO is designed so that it can cover the desired 20 MHz to 30 MHz range with the range of DC voltages that can be produced by the phase/frequency detector. The channels are selected by changing the value of N in the programmable frequency divider. Once the loop has "achieved lock", the frequency out of the programmable divider will be equal to 10 KHz.

A phase detector can be made using an exclusive or gate: a frequency detector can be made using flip-flops and gates. To make it easy, you can buy several different commercial IC's (such as Motorola's 4044) which includes all of this circuitry as well as a VCO.

The loop filter is a low pass filter which will attenuate the 10 KHz reference frequency. If the filter's break frequency is set too low, the loop will take an excessive time to reach lock. Design of the filter is important. It affects the loop stability, lock-up time, and residual FM. Since the VCO's frequency (and hence its phase) is controlled by a DC voltage, just a small amount of noise on the control line can cause unwanted FM or noise. The Q of the components in the VCO affects the close-in noise spectrum of the PLL.

A small amount of audio can be injected in to the output of the loop filter to create an FM signal. If the reference signal itself is an FM signal, demodulated audio can be recovered out of the loop filter. Both of these approaches are shown here:

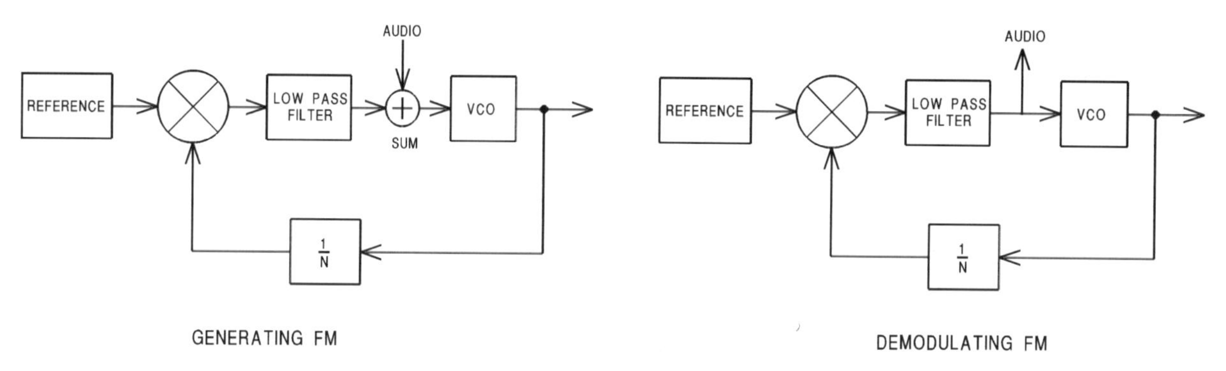

Although PLL's are a very powerful tool, it is easy to get into trouble if you are trying to achieve good spectral purity. The prospective designer is strongly urged to read the multitude of articles and texts that cover PLL design before plunging ahead with any serious designs.

FREQUENCY SYNTHESIS

Modern transceivers all use frequency synthesizers to generate both the transmitted RF frequency, and the local oscillator frequencies for the receiver. HF ham transceivers built prior to about 1975 did not use synthesizers: a stable VFO was used in combination with a number of switcheable, crystal controlled local oscillators. Unfortunately, VFO's are not as stable as you might like, and their tuning rate is not completely linear. Transceivers had to include crystal-controlled calibrators so that the operators knew "where they were" in the band. Modern synthesized rigs are very stable, provide precise frequency readout, and are able to offer all kinds of memory-based features.

Synthesizers have been used in VHF equipment since the late 60's. Since these frequencies are typically channelized, it was natural to use PLL-type approaches. For HF communication, users expect to have the "feel" of a VFO-tuning dial, so it took a while before suitable rigs were developed.

Most modern HF transceivers use weighted flywheels on their tuning knobs, and quadrature optical interrupters to detect the motion of the dial. As the dial is turned, optical pulses cause binary up/down counters to change their value. The contents of these counters is what determines the frequency, and what is displayed on the read-out. To "feel" like an analog VFO, the tuning steps must be less than 50 Hz each: many rigs use 10 Hz or even 1 Hz. Therefore, you can consider a modern 0.1 MHz to 30 MHz transceiver as having 2,990,000 channels that are spaced 10 Hz apart!

Phase-Locked Loop Frequency Synthesizers

We have already discussed the use of PLL's to generate stable, frequency-locked signals. If a single PLL were to be used on an HF transceiver, the reference frequency would have to be 10 Hz. With this low a reference frequency, the constants of the loop filter are such that the lock-up time would restrict the operator to tuning very slowly: perhaps no more than 200 Hz per second! This is completely unusable, so many modern transceivers use multiple PLL's with higher reference frequencies. Mixers and filters are used to effectively combine the outputs of the PLL's. The result is an overall design that still has an effective 10 Hz channel spacing, but the short lock time on the multiple loops allows the tuning dial to be "spun" just like on the good old analogue VFO's! A block diagram indicating how a triple conversion HF SSB receiver could be designed using multiple PLL's is as follows:

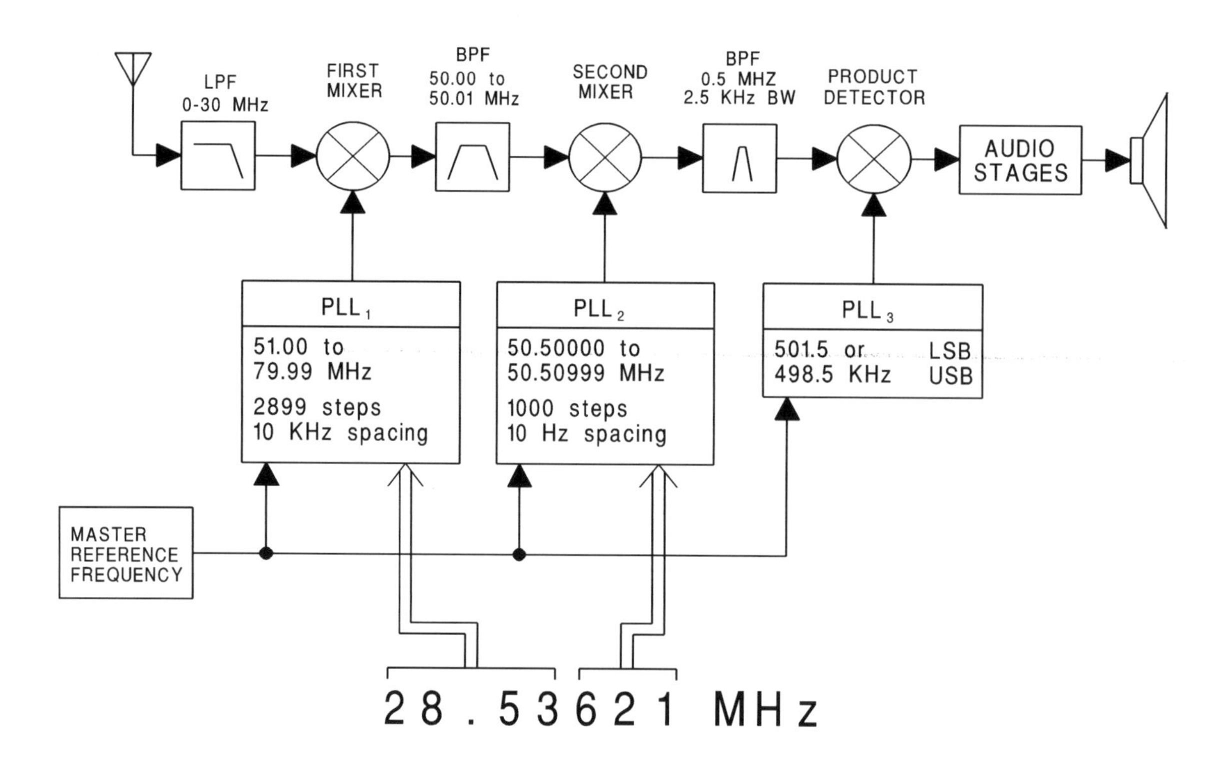

This block diagram is meant to represent a hypothetical design - it does not match any known commercial receiver. The first IF frequency is not completely constant - it varies over a range of 9.99 KHz as PLL_2 is tuned. Note that the tuning dial will operate both PLL_1 and PLL_2. In order to keep the lock-up time short on each PLL, the reference frequency into the phase/frequency detector must be kept high. One way that our hypothetical HF receiver could implement PLL_1 is as follows:

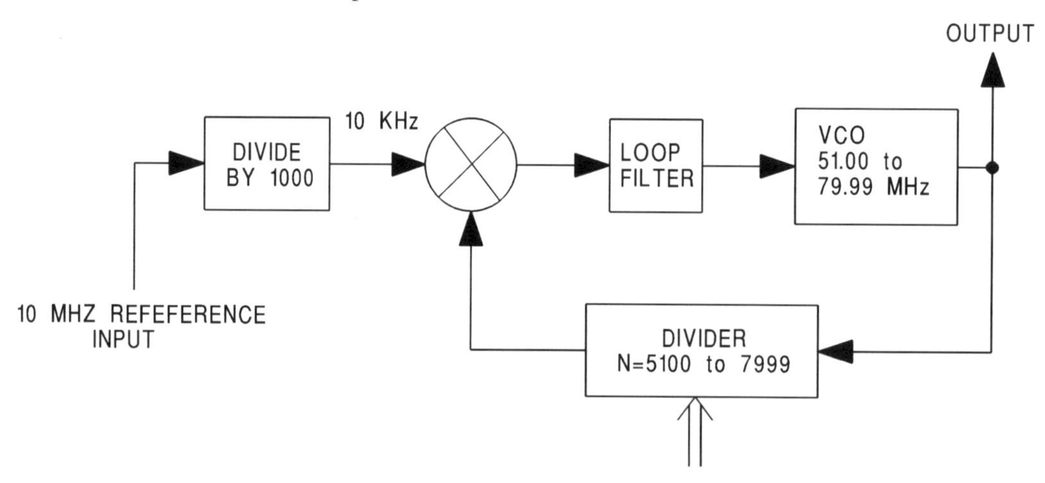

This design uses a reference of 10 KHz, so lock in time will be moderately fast. If the user "spins" the tuning knob, this loop only needs to re-establish lock if the dial has changed by more than 10 KHz. The design of PLL_2 needs to be more complex in order to allow 10 Hz effective output channel spacing while working with a reasonably high reference frequency. This is done using a combination of mixers, filters, and dividers:

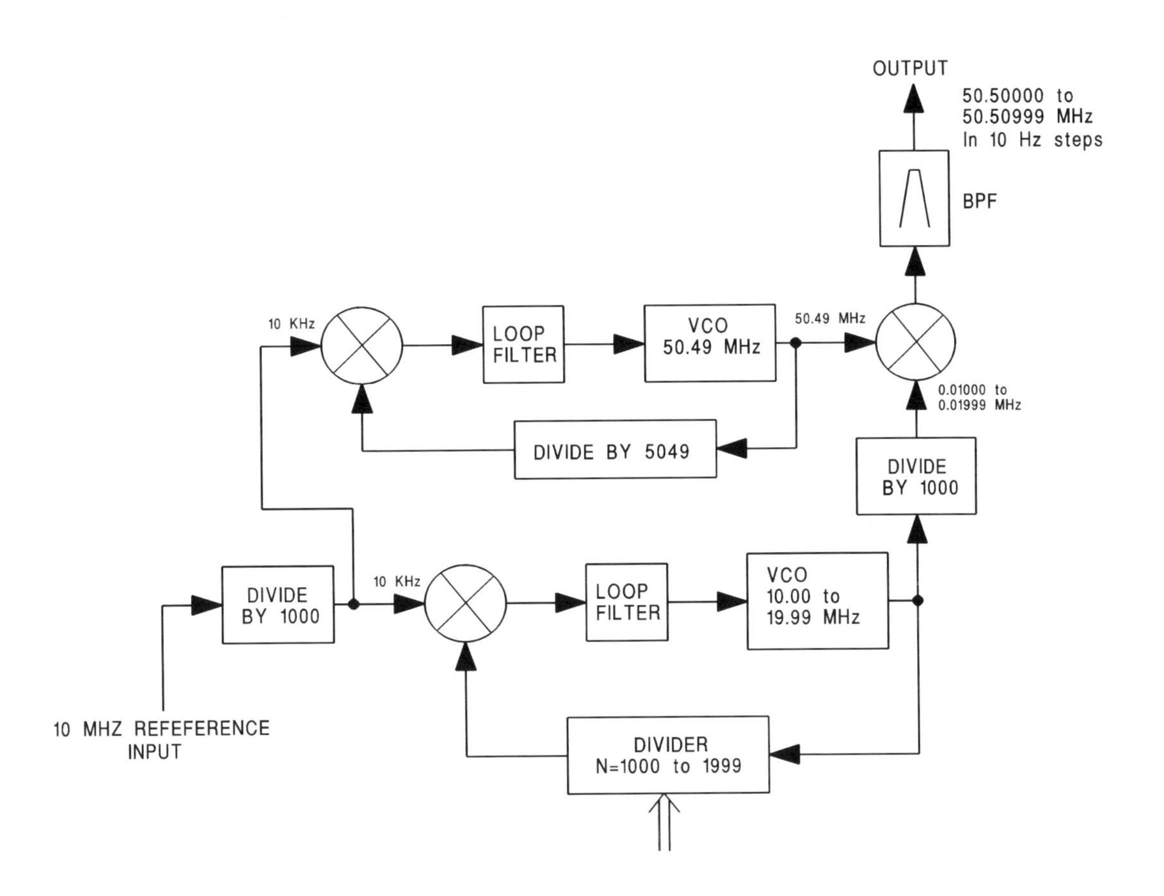

The reference frequency of the main PLL in this circuit is 10 KHz, so the lock-in time is still reasonable, even though the output channel spacing is 10 Hz. The secondary PLL in this circuit operates at a fixed frequency of 50.49 MHz, so lock-in time is not a consideration.

Stand-alone laboratory frequency synthesizers are available from companies like HP, Fluke, Gigatronix, Rohde & Schwartz, or Watkins-Johnson. These units have wide coverage (up to well over 1 GHz), high resolution (1 Hz or better), and exceptional frequency stability. They also cost more than many automobiles! One of the important parameters that has a large effect on the price you pay is the **phase noise** spec. Phase noise relates to the spectral purity of the output signal.

Ideally, a synthesizer has all of its output energy concentrated at a single frequency, and none anywhere else. Any synthesizer or oscillator has a distribution of energy about the central output frequency. If these "tails" do not decrease rapidly, it will not be possible to use a synthesizer for adjacent channel testing of receivers. If the synthesizer in a receiver also has significant "spectral tails", the receiver's overall noise level will appear to be artificially high, and the close-in signal rejection may be less than optimum. This is why some older VFO-based receivers sometimes seem to have less background noise than some synthesized radios.

Phase noise specs are usually stated as SSB Phase Noise in dB per Hertz relative to the carrier, at an offset of 20 KHz from the carrier. A good spec would be -100 dB, a great spec would be -130 dB.

Direct And Indirect Frequency Synthesis

Apart from using a PLL, there are two other approaches to frequency synthesis - Direct Synthesis and Direct Digital Synthesis (DDS). Both can give extremely fast frequency switching times. In a direct synthesis unit, a stable frequency reference is fed through multipliers to generate the desired output frequency. A simple approach that shows how the integral frequencies from 1 to 10 MHz could be generated by multiplying a 1 MHz reference is shown below:

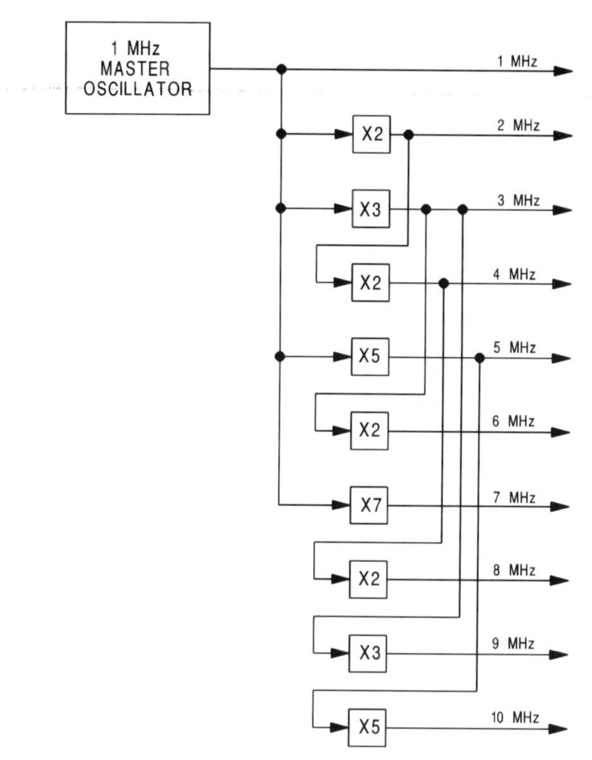

If mixers and sideband filters are used as well, it is possible to produce non-harmonically related frequencies. One system that uses this approach is called "Iterative Direct Synthesis", which we will abbreviate as "IDS". This was originally offered in a line of products by General Radio. A block diagram of part of a hypothetical IDS synthesizer is reproduced below:

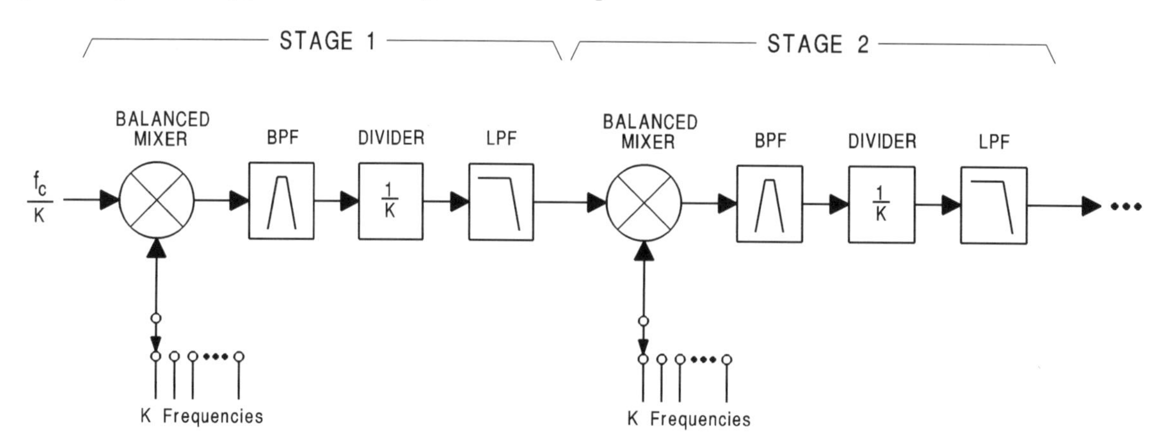

The K reference frequencies and f_C are all generated directly from a single stable reference oscillator by means of multipliers, dividers, mixers, and filters. The selection switches shown in the block diagram are actually transmission gates or diode switches that are controlled from the synthesizer's front panel controls. This synthesizer design offers very good phase noise specs and almost instantaneous frequency switching. Stages are cascaded to achieve the overall resolution required. Although the performance is good, the complexity, size, and cost are high.

Direct Digital Synthesizers

A more recent development is the Direct Digital Synthesizer, or "DDS" approach. This uses straight forward "brute force" digital techniques. Because digital integration is becoming more and more dense, the costs and practicality of this approach become more reasonable all the time. The basic approach relies on a finding by Nyquist many years ago: a signal can be re-created from samples, as long as there are at least two samples for each cycle of the highest frequency to be reproduced.

In a DDS synthesizer, a binary counter referred to as the phase accumulator is connected to a read-only memory (ROM) that contains a stored word representing the value of the trigonometric sine function for that particular angle. The output of the ROM is then converted to an analog voltage by a very high speed D/A converter. In other words, if the phase accumulator is thought of as holding a binary representation of the instantaneous phase of a sine wave, the output of the D/A converter will be an analog voltage whose value is proportional to the sine of the phase angle. Using a very high speed clock, values are added to the phase accumulator on a repetitive basis. The net effect is that then the output voltage from the D/A converter will start to produce a sine wave whose frequency is proportional to the value that is added to the phase accumulator on each cycle of the clock. The output of the D/A converter is passed through a low pass filter to remove traces of the reference clock and switching artifacts. There are no loops to lock, so response is effectively instantaneous. The block diagram on the following page shows the general arrangement of o DDS synthesizer.

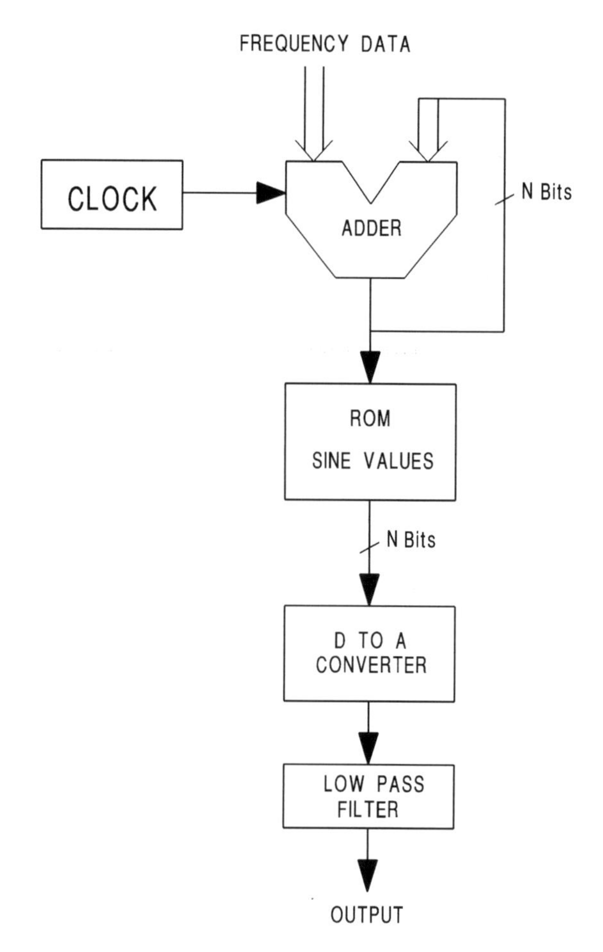

The clock frequency must be at least twice the highest output frequency to be produced. The block labeled as "adder" consists of digital circuitry that is capable of adding two binary words. The output of the adder implements the phase accumulator. At every cycle of the clock, the adder adds the value from the frequency programming data on to the previous value of the phase accumulator - if the unit is set to generate a low frequency, the phase accumulator will increment by only small amounts each cycle.

In order to offer decent spectral purity, the D/A converter and phase accumulator should have many bits of resolution, and a high reference clock speed should be used. Modulation can be done completely in the digital domain. In-phase and quadrature (I & Q) outputs can be simultaneously produced. Several semiconductor companies offer major DDS building blocks as integrated circuits. Analog Devices is the most prolific. Maximum output frequency is up to 1000 MHz.

Modern "Hi-End" transceivers often use a combination of PLL and DDS technology in order to achieve high resolution, fast lock-in time, and excellent spectral purity.

VOICE MODULATION AND DEMODULATION

The whole purpose of a radio system is to convey information. In order to accomplish this, the basic RF carrier has to be **modulated** with the information to be conveyed. In this section we will primarily be looking at ways of transmitting analogue voice data. Looking first of all at simple analog audio signals, there are three classic ways of modulating a carrier: amplitude modulation (AM), phase modulation (PM), and frequency modulation (FM). Specialized forms of AM include single sideband (SSB), double sideband suppressed carrier (DSB), vestigial sideband (VSB), and amplitude companded single sideband (ACSB).

Amplitude Modulation

In AM, the amplitude of a carrier is multiplied by the modulating signal plus a constant value. If A is the peak carrier amplitude (single-sided), and B is the peak amplitude of the modulating signal, the output of the modulator will be: $A[\sin(f_c)] \cdot B[1+\sin(f_m)]$. As we discussed in the section on mixers, this multiplication will result in three output frequencies: the original carrier and two sidebands, each separated from the carrier by the modulating frequency f_m. The output of the AM modulator will therefore be -

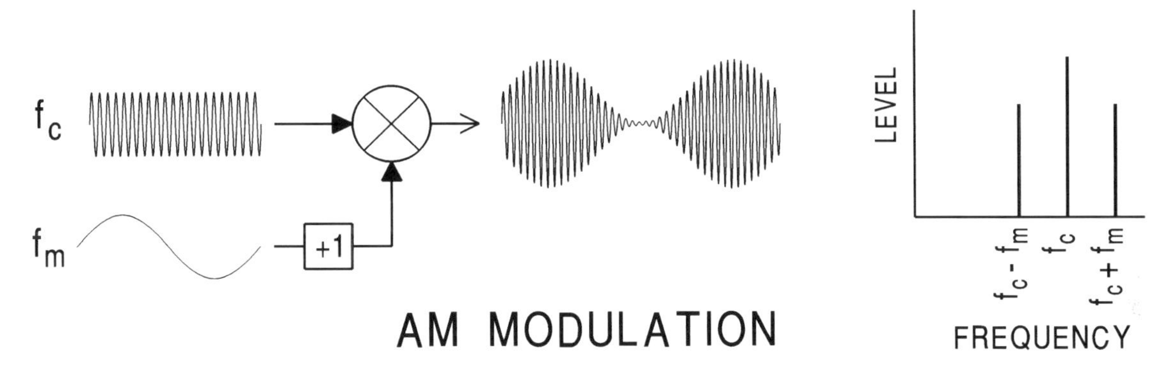

AM can be demodulated by a simple diode detector and a low pass filter - this is called "envelope detection". It can also be demodulated by a product detector (a balanced mixer) if the L.O. is synchronized with the carrier - this is called "coherent detection". AM is very wasteful of spectrum. If normal communications quality voice components are from 300 to 2300 Hz, the AM signal will occupy 4.6 KHz of RF spectrum. The carrier is constant: it consumes power whether or not information is being conveyed - this is not desirable for battery-powered transmitters.

Double Sideband

DSB also involves a multiplication process, but the carrier component disappears. The output in the time domain is $A[\sin(f_c)] \cdot B\sin(f_m)$. The output spectrum is identical to that of AM, except that the carrier is not present. The DSB signal will still occupy a total of 4.6 KHz of spectrum. Because no RF power is being transmitted in the absence of voice, this technique is more power-efficient than AM.

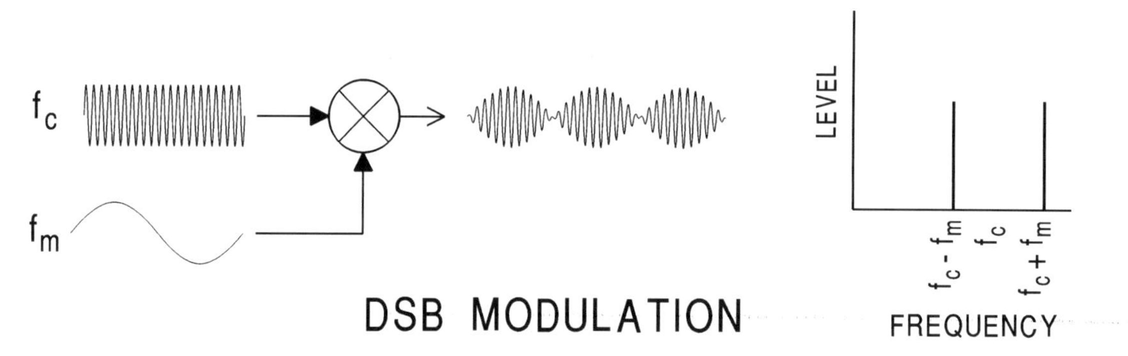

Note that in a DSB output, the phase of the RF carrier actually undergoes a 180 degree reversal at the zero-crossing points of the envelope.

Single Sideband

SSB is generated in a similar fashion to DSB, except that one sideband is removed either by filtering (most common) or by cancellation (the phasing method). The RF bandwidth occupied is only 2 KHz (assuming that the range of audio frequencies to be transmitted covers the range of 300 Hz to 2.3 KHz). A comparison of the RF spectrum usage of the three amplitude modulation schemes is as follows:

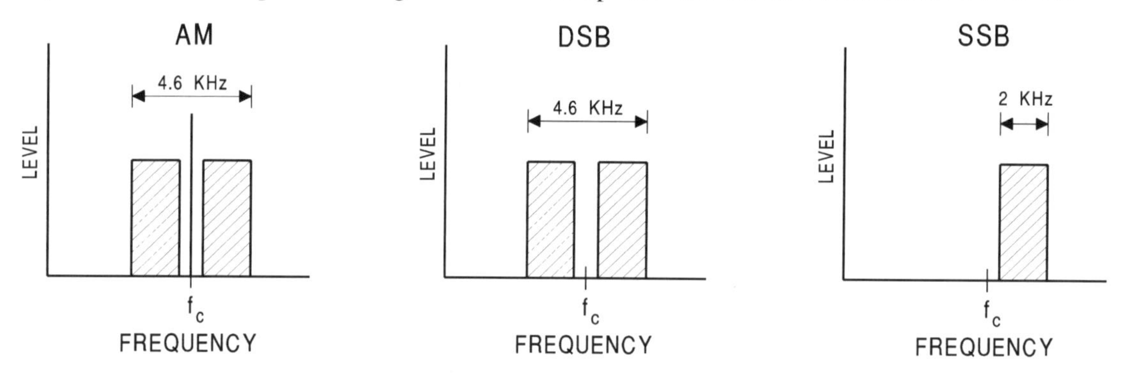

Both SSB and DSB are demodulated by a product detector (basically a balanced mixer) with an L.O. frequency that matches the suppressed carrier. If an AM signal is carefully tuned in (zero-beated) with an SSB receiver utilizing good filters, it is possible to independently demodulate the upper and lower sideband. It is theoretically possible to put different information channels into the two sidebands and then demodulate them separately using this **independent sideband** receiver.

VSB is just like AM, except that one of the sidebands is <u>mostly</u> removed with a filter. This is the approach used in the NTSC television system. Bandwidth is reduced over AM, and a simple diode detector can be used for demodulation, but the carrier power is still constant.

Frequency Modulation

FM enjoyed much popularity in the 50's because it did not require that the receiver be perfectly "tuned in". Non-crystalized VHF and UHF receivers tended to drift around quite badly, making even AM, let alone SSB difficult to receive without constant attention. Commercial broadcasters liked FM because, once the signal strength exceeded a certain threshold, very good S/N performance was possible, thereby offering the perception of good audio fidelity. PM is very much like FM, because the integration of a phase change over time is equivalent to a change in frequency. At the fundamental level, PM and FM can be created from each other by simple RC networks in the modulator and demodulator. We will look just at FM.

In an FM modulator, the instantaneous frequency of the carrier is proportional to the instantaneous voltage of the modulating signal. The simplest way to generate FM is to apply audio to the control input of a voltage-controlled oscillator (VCO). This is how FM is generated in most synthesized FM radios. The output level of an FM signal is constant, and a carrier is transmitted even in the absence of any audio modulation: FM is not as power efficient as SSB for voice communications.

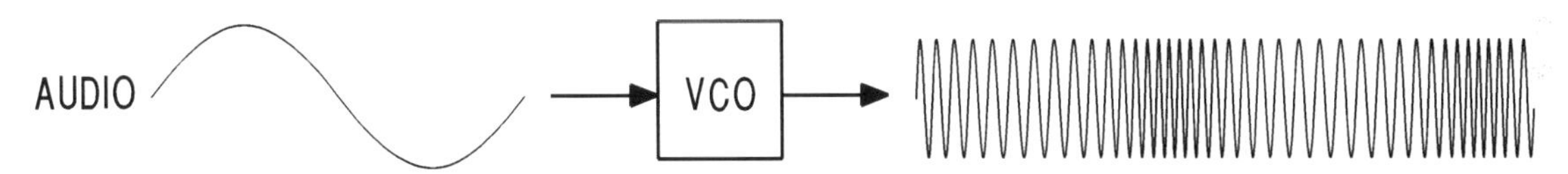

The mathematical analysis of a frequency modulated signal gets rather complex. We will need to treat narrow band FM systems separately from wide band FM. The terms "wideband FM" and "narrow band FM" are relative to the amount of effect the modulating signal has on the instantaneous carrier frequency. A term called **modulation index** (the symbol "β" is used to signify this term) defines the ratio of the peak deviation of the carrier frequency (Δf) by the peak modulating frequency. Assuming that the amplitude of all modulating signal frequencies is equal, this suggests therefore that β will decrease linearly for higher modulating frequencies. Most FM systems use a pre-emphasis circuit in the transmitter: this is nothing more than an RC high-pass filter which accentuates the amplitude of the high frequency modulating signals: it is balanced by a de-emphasis circuit in the receiver - the effect is to boost the deviation for higher frequencies, thereby maintaining a more constant β.

A narrow band FM (abbreviated as NBFM) signal uses a β of no more than approximately 0.5. A narrow band FM system has similarities with AM: there is a fairly constant carrier, together with sidebands above and below the carrier, separated from the carrier by the modulating frequency. The only difference is that the sidebands are at an angle of 90 degrees with respect to the sidebands of a corresponding AM system. A narrow band signal can be generated by modulating a varicap in the primary frequency determining circuit, or by injecting the audio signal onto the VCO's control pin if the transmitter uses a PLL for frequency generation, or by using the following approach:

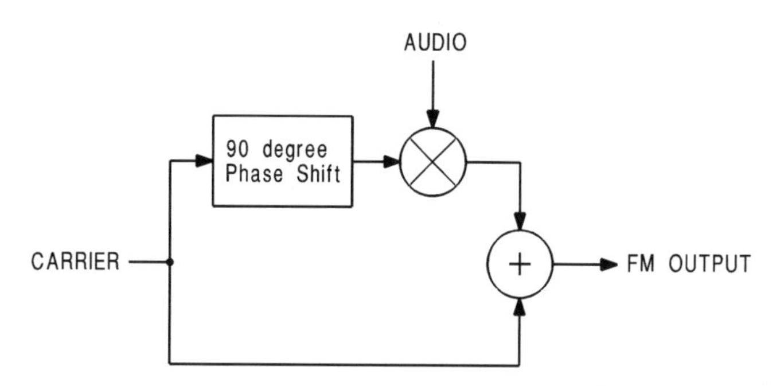

An NBFM spectrum consists of a carrier together with a group of upper and lower sidebands. The sidebands are separated from each other by the modulating frequency. The amplitude of the sidebands decreases as they get further away from the carrier frequency. In the narrow band case, the amplitude of the sidebands falls off so quickly that we are only primarily concerned with the first ones, and the total RF bandwidth can be considered as being approximately equal to that of an AM signal. The sum of the power in the carrier and the sidebands must be constant, so the energy at the carrier frequency decreases as energy is put into the sidebands.

Commercial FM broadcasting (88 MHz to 108 MHz) uses wideband FM (WBFM). The maximum deviation (Δf) is about 75 KHz, and the maximum modulating signal frequency (other than the stereo subcarrier) is about 15 KHz, therefore β is approximately 5. For this large a value of β, the sidebands do not fall off very quickly as you move away from the carrier. Significant energy is going into the sidebands, and the centre frequency amplitude decreases accordingly. At certain modulating frequencies and deviations, the carrier appears to disappear completely. The bandwidth requirements are much greater for WBFM than for NBFM. As a rough approximation, for very large values of β, the total RF bandwidth required is equal to $2\beta f_m$ or twice the deviation. An approximation to the bandwidth required for an FM transmission is given by **Carson's Rule**, which is:

$$\text{Bandwidth} \cong 2\Delta f + 2f_m = 2\Delta f\left(1 + \frac{1}{\beta}\right)$$

In theory an FM signal has an infinite number of sidebands. If we ignore the sidebands that together comprise less than 1% of the total signal power, we arrive at the following chart for the required transmission bandwidth as a function of Δf and β:

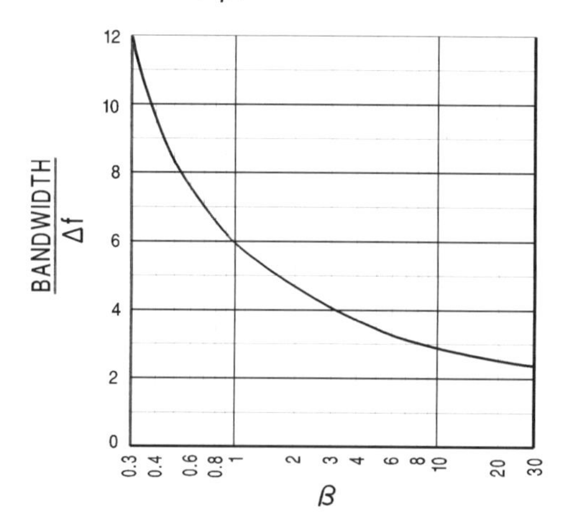

This chart should only be used for WBFM. If the bandwidth of a receiver is set equal to that indicated on the above chart, 99% of the total transmitted power will be received. In practice it is possible to use somewhat less bandwidth than that stipulated by the chart. Note that at very large values of β, the curve approaches a value of $B/\Delta f = 2$.

Most VHF and UHF transmitters contain frequency multipliers in the RF chain. As an FM signal passes through a multiplier, the deviation (and hence the modulation index β) increase by the multiplication ratio. Conversely, passing an FM signal through a frequency divider will decrease the deviation by the division ratio.

The information in an FM signal is contained in the instantaneous frequency not the amplitude. FM receivers often contain limiters, which are basically saturated amplifiers which remove all amplitude variations. FM is particularly immune from impulse noise interference if the received signal amplitude is above a specific threshold value.

An FM signal can be demodulated using a PLL as discussed earlier. It can also be demodulated by a **discriminator**, which is a circuit which has a linear and symmetrical voltage versus frequency response. NBFM can also be demodulated by slightly miss-tuning an AM receiver and using "slope detection", although many of FM's noise advantages will be lost. If an FM signal is down converted to a low frequency (such as 100 KHz) and turned into a square wave (by using a comparator), it can be demodulated by using a "pulse counting discriminator", which is nothing more than a "one-shot" and a low pass filter.

Modulation Comparison

It is useful to compare the relative performance of AM, SSB, and FM. A good indication of the relative merits of the different modulation systems is to examine the signal to signal-plus-noise ratio at the output of the demodulator, as a function of the S/N ratio of the communication channel.

Any communication channel has random, broadband noise. This has uniform power per unit bandwidth, and is so-called **white noise**. The power is measured in units of "Watts per root Hertz", and the amplitude is in "Volts per Hertz". On the communication channel, we will call S_c the power of the signal, and N_c the power of the noise in the system's bandwidth. Thus the signal to noise ratio going in to the receiver is simply S_c/N_c. Note that N_c is proportional to the receiver's bandwidth, which in the following discussion we will set equal to that required for the particular form of modulation used.

We will assume for the moment that the receiver does not add any noise until the signal is finally applied to the demodulator. We are now interested in knowing what the signal to noise ratio S_o/N_o at the output of the demodulator is relative to the input to the receiver S_c/N_c.

For an AM receiver using coherent detection:

$$\left(\frac{SNR_O}{SNR_C}\right)_{AM,Coherent} = m^2$$

The value of m is the modulation percentage divided by 100. It can be seen therefore, that for a 100% modulated AM signal, the S/N ratio at the output of the detector is the same as the S/N ratio of the channel. If the percentage of modulation is less than 100, the output S/N ratio will be worse than that present on the channel.

In practice, most AM receivers use an envelope detector. This will result in a somewhat lower output SNR:

$$\left(\frac{SNR_O}{SNR_C}\right)_{AM,Envelope} = \frac{m^2}{2+m^2}$$

For 100% modulation, m=1 and the S/N ratio at the output of an AM receiver equipped with an envelope detector will be one third that of the input channel.

For an SSB receiver,

$$\left(\frac{SNR_O}{SNR_C}\right)_{SSB} = 1$$

For an FM receiver operating at signal levels in excess of the threshold level,

$$\left(\frac{SNR_O}{SNR_C}\right)_{FM} = \frac{3\beta^2}{2}$$

This indicates that the output S/N from an FM system will be $1.5\beta^2$ higher than a 100% modulated AM system (using coherent detection), given the same starting channel S/N ratio. Therefore, with large deviations (and therefore large values of β), FM offers substantially higher S/N ratio than a corresponding AM system. This was one of the major appealing factors to broadcasters who wanted to use FM to transmit high fidelity programs. However, high values of β also require large amounts of channel bandwidth (about 240 KHz for a broadcast FM station). This trade off of channel bandwidth versus output S/N ratio is a characteristic of all noise-improvement systems. However, the above equation assumes fairly large values of S/N ratio in the channel: not "weak signal" operation! As β is increased, the receiver bandwidth must be increased, thereby collecting more broadband noise power: at some point, the effective input S/N ratio is such that the assumptions used in generating the above equation fall apart, and the noise takes over.

FM systems are often designed with pre-emphasis circuits in the transmit audio stages, and corresponding de-emphasis circuitry in the receive audio sections. A pre-emphasis circuit is simply an RC high pass filter, while the de-emphasis circuit is a low pass filter. The idea is to maintain a more constant value of β for varying audio frequencies.

Assuming that the receiver input signal level of an FM system is well above the threshold, an improvement of average output S/N ratio of several dB is possible. This approach is used in broadcast WBFM transmissions, where an overall output S/N improvement of over 6 dB is achieved.

FM has a pronounced **threshold effect**. For input S/N ratios which are higher than the threshold, the noise improvement ratios predicted by the above equation for WBFM are true. Below the threshold, the output S/N ratio deteriorates rapidly. The actual level of the threshold S/N ratio depends on β. For large values of β, the threshold occurs at an input S/N ratio of about 13 dB. For values of β which are below approximately 0.5 (note that this defines NBFM), there is no improvement of output S/N ratio over AM.

Experimental studies have confirmed this data. The following chart clearly shows both the threshold effect, and the S/N improvements for FM systems with larger values of β and amplitudes that are above the threshold:

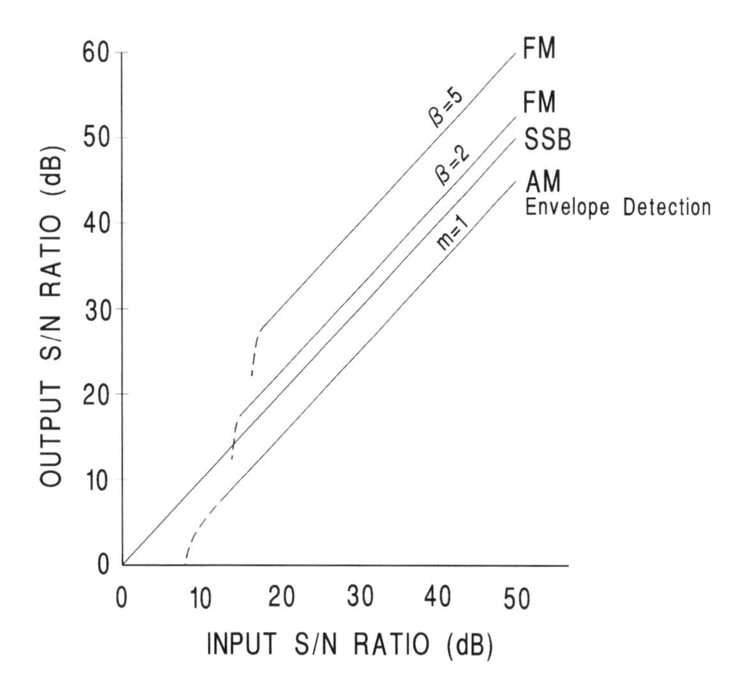

Below 450 MHz, all amateur FM transmissions are narrow band. The S/N ratios for FM are therefore not much better than AM. For mobile operations, where the actual amplitude of the received signal is rapidly changing, FM offers the advantage of being amplitude-independent.

Weak signal voice communication is primarily done with SSB. All the transmitted power goes into producing the sideband that contains the information, and the required bandwidth is less than half that of an AM or NBFM signal. The above chart assumed that the receiver bandwidth was sufficient to pass the signal's required sidebands. SSB is the narrowest bandwidth analogue modulation system, and it will therefore have the largest input S/N ratio (assuming constant transmitter power and uniform noise spectrum). This re-confirms that SSB is the preferred technique for weak signal voice transmission.

The book "Single Sideband Systems and Circuits" by Sabin and Schoenike discusses a term called the **articulation index** (AI). This attempts to measure the intelligibility of the human voice over a communication channel in terms of the AI, which can have a value between 0 and 1. In order to achieve a value approaching 1, a channel would have to have virtually infinite S/N ratio, and the system would have to pass audio frequencies of over 6 KHz. Trained operators using a limited vocabulary (in other words, patient "hams" who are trying to "pull out" a weak signal) can operate (with fatigue) using an AI as low as 0.2. A value of 0.3 provides useable communication (with some fatigue), while a value of 0.5 gives comfortable copy.

The following chart gives a very interesting comparison of AM, FM, and SSB in terms of their AI vs input S/N ratio. This chart assumes that the audio bandwidth is limited to the range of 300 to 3000 Hz, and has no clipping or pre-emphasis. The receiver bandwidth is just wide enough for the signal in each case.

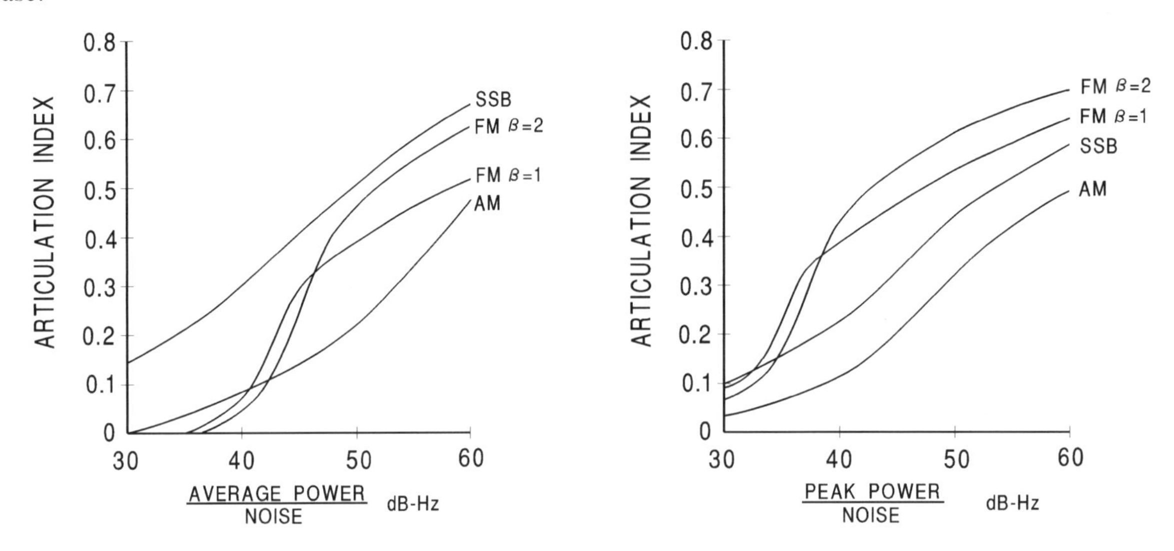

This chart suggests that based on average power ratings, SSB has about a 15 dB advantage over AM, and a 7 dB advantage over NBFM. If the transmitter is peak power limited, FM is preferred.

All of the above discussions have been focused on the transmission of voice signals using analog modulation techniques. It is obviously possible to sample and digitize an analog voice signal to produce a stream of digital values that can then be transmitted via digital techniques. If the audio passband is limited to about 3 KHz, the minimum practical sampling rate is 8 KHz. For decent fidelity and dynamic range, we might wish to digitize to a 6 bit resolution: this would therefore require a data transmission rate of 48 Kbps. This sounds like it will occupy a large bandwidth! However, we can use non-linear digitizing to reduce the sample resolution to perhaps 4 bits, and then we can step into the very complicated world of **vocoders** and perhaps reduce the overall data rate down to below 4 or 5 KHz. Several vocoders have been developed, but the one which is the most commonly adopted (because of its efficiency) is the AMBE vocoder. For the moment, how this is done is beyond the scope of this text!

DATA MODULATION AND DEMODULATION

In data transmission, we are concerned with the transmission of 1's and 0's. Rather than talk about the S/N ratio of the output, instead we are concerned with the **Bit Error Rate** (BER). For a given channel BER, it is of course possible to add additional information to the data to allow the receiver to detect when bit errors occur, and then arrange to have the unknown data re-transmitted. It is also possible to add addition error correcting characters (ECC) which can be used at the receiver to reconstruct the original digital message, even if bit errors have occurred. Typically, it takes approximately two ECC characters to correct a single data error, so there is a limit on how much overhead you want to add to the data vs the acceptable overall BER. As ECC characters are added to a message, the number of bits to be transmitted increases, and therefore the required channel bandwidth increases if the message is to be transmitted in the same length of time. Hopefully you will recall from the previous discussions that WBFM also had a similar trade-off of channel bandwidth vs output S/N ratio.

Conventional analog radios designed primarily for voice (like most of today's amateur gear) can be used to convey digital data through the use of a modem. A transmitting modem takes digital data in and translates the information into a form that is capable of passing through the analog circuitry in the voice-optimized equipment.

It is important to understand the distinction between the **data rate** and the **signaling rate**. The signaling rate is the rate (measured in changes per second, or **baud**) that the output of the modem changes. The data rate is the rate at which the actual digital data is transmitted in bits per second (bps). The simplest systems use identical data rate and signaling rate, but there may be a large difference between the two rates in more sophisticated equipment.

The data bit stream coming from the source could obviously be used to modulate the amplitude of a carrier: this is completely analogous to CW. However, the BER of the overall system would be poor due to interference from impulse-type noise and rapidly varying signal strengths. Most systems involve changing phase or frequency.

In frequency shift keying (FSK), the carrier's output frequency varies between two values, depending on whether the data is a 1 or a 0. If an FSK signal is tuned in on an SSB receiver, the output from the detector will be an audio tone which switches between two frequencies. This can be detected using two narrow band audio filters, or via a PLL.

Another approach uses the digital data to FSK an audio oscillator: the output can then be put into the speech channel of a transmitter (AM or FM), and the audio output from the receiver can then be demodulated, again using two audio filters or a PLL. This is called AFSK, and has the advantage that miss-tuning the receiver does not affect the frequencies of the two tones. If an AFSK tone is used to modulate an SSB transmitter however, the output is again straight FSK.

Both FSK and AFSK have identical signaling and data rates. Because of this, and the fact that the signals need to be compatible with equipment designed primarily to pass audio frequencies of up to 2.5 KHz, the data rate is usually no more than 1200 bps.

At frequencies of 430 MHz or above, the amateur regulations allow the transmitted bandwidth to be up to 12 MHz wide. If a custom-built FSK transmitter and receiver were built, it would be possible to transmit very high data rates using a simple FSK scheme. Unfortunately, a great deal of effort is required, and you probably wouldn't have anyone to talk to other than yourself! More sophisticated modulation schemes are necessary if you wish to transmit high speed data without consuming lots of the RF spectrum.

The usage of RF spectrum is primarily due to the rate at which the output signal from the transmitter changes. In other word, it is a function of the baud rate, **not** the data rate. Knowing this, we can cook up schemes to cause the output of the transmitter to change states at a slow rate, but convey more than one bit of information in each "state". This sounds like magic, but it isn't really that hard to think of ways of doing this. Think of the data that needs to be transmitted as a big long string of bits. Now, instead of transmitting 1 bit at a time, why not transmit two bits at a time? Let's look at an example.

We will break up the transmitted data into "pairs of bits", or "dibits". Each dibit can have one of four possible values: 00, 01, 10, and 11. Now, let's modulate the phase of a transmitter's carrier between four possible values: 0°, +90°, -90°, and -180° depending on the value of the dibit to be transmitted. Now, if we cause the phase changes to occur at 2400 transitions per second (referred to as 2400 baud), the actual rate that data is being transmitted is 4800 bits per second! The demodulation can be done with a PLL and some other logic components. This scheme is called Quadriphase Shift Keying (QPSK).

Well, if we were able to double the transmitted data rate without changing the signaling rate using this approach, why not try handling more than two bits at once? If three bits are transmitted at once by using 8 possible carrier phases, we can have a data rate that is three times as high as the signaling rate: this is called 8PSK. This scheme can be pushed to 16PSK before tradeoffs start to rain on the party. The following diagram represents the possible output states of the carrier of a transmitter using 4 QPSK and 16 QPSK. Each dot represents one output state. Distance from the centre represents amplitude, and angle represents phase. These are referred to as **signal constellations**:

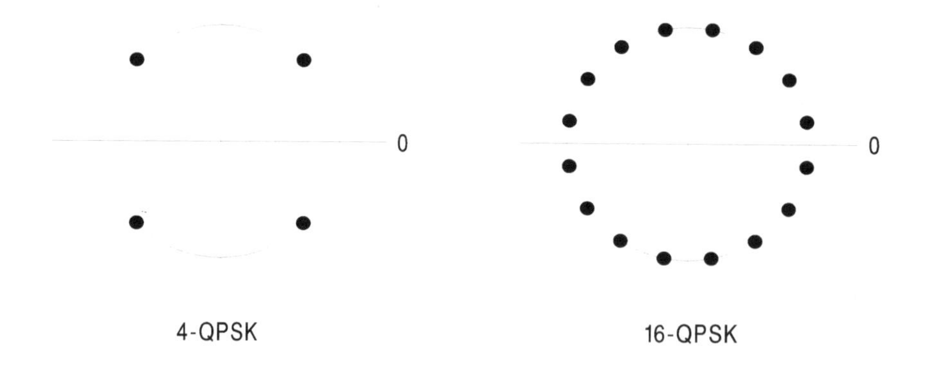

Rather than modulating the carrier's phase, it might be possible to vary the frequency or amplitude instead. Amplitude modulation is not normally used by itself, because of reasons already discussed. Frequency modulation by itself could be used, but most modern high speed modems use a combination of phase and amplitude modulation. We will now look at an example where the data rate is 4 times the signaling rate.

A system called 16QAM (for "16 state quadrature amplitude modulation") transmits 4 bits at a time. The 4 bits are split into two dibits, one of which can be considered as controlling the carrier's amplitude, and one of which controls the carrier's phase. The end result is that there are 16 unique combinations of carrier phase and amplitude. The scheme has been extended in some high speed satellite communication systems to 256-QAM. The signal constellations for the two systems are shown below:

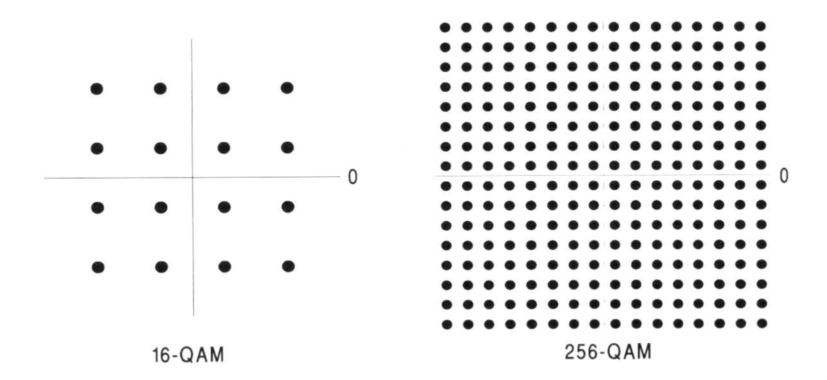

The modulation and demodulation hardware involves the use of quadrature modulators, which are best understood as a part of studying **complex** signals (which have real and imaginary components). For now, we will just assume that this is possible, and worry about the details later. The receiver's demodulator has to differentiate between the different unique states of the carrier. It should be intuitively obvious that in the presence of noise it will have more difficulty differentiating states in a 256-QAM system than a 16-QAM system.

The objective of all digital communication systems is to recover the original transmitted sequence of 1's and 0's. At some point in the receiver, a circuit is used to decide whether it is currently receiving a 1 or a 0. Timing recovery circuitry has presumably generated an internal clock that is synchronized with the transmitter's clock, and this is used to sample the received analogue waveform at the best sampling time, which is in the middle of a transmitted state. The output of the sampler is then compared to a reference voltage in a comparator circuit in order to generate a digital 1 or 0. In order to minimize false decisions caused by random noise that may effect the received voltage right at the sampling time, some type of filter is used just before the sampler and comparator. If the signaling rate is low and the SNR is high, a simple low pass filter will be sufficient to give a very low error rate.

Multi-path distortion and broadband noise will contaminate the received analogue waveform as it is fed to the sampler and comparator. The finite bandwidth of the receiver will also affect the rate at which the analogue waveform changes its state before being sampled, and intersymbol interference will result. The net effect of all of these considerations is that the waveform being fed to the sampler and comparator will become less concise, and the individual states will become smeared, making it difficult to extract the correct data. One way of evaluating the quality of the signal before it is sampled and compared is to use an **eye pattern.** A storage oscilloscope records multiple waveforms at the input to the sampler as the waveform goes through a sequence of transitions. The oscilloscope's timebase is synchronized to the system clock. The height of the opening in the eye pattern at a particular sampling time gives an indication of the noise margin.

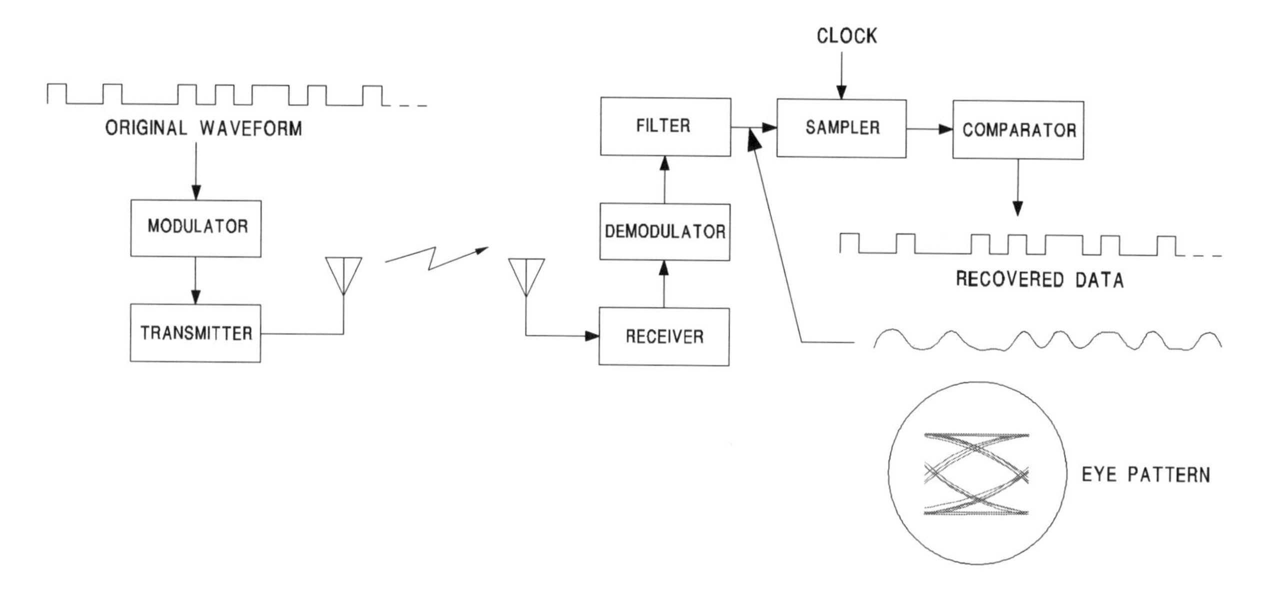

One disadvantage of using large numbers of states is that (as can be seen by looking at an eye pattern) it gets harder and harder to differentiate between one state and the next in the presence of noise. This must be offset either by ensuring a higher channel S/N ratio, or by using more and more error correction (which itself consumes part of the channel bandwidth). As you can tell, this is not a trivial topic!

SPREAD SPECTRUM

Most hams find it hard to think of a modulation scheme that intentionally uses large amounts of the RF spectrum as being an "advancement": most of us (especially on HF) are concerned with keeping the bandwidth as small as possible while maintaining effective communication. Spread spectrum is a modulation method that trades off bandwidth for spectral power density, and it offers some surprising performance advantages for data transmission.

Spread spectrum (abbreviated SS) systems have been around in one form or another for over 70 years. They were originally designed to allow military communication to occur in the presence of intentional jamming, but then developed as viable communication schemes based on their own merits. Today, a variety of commercial digital communication systems (including the GPS navigation system and most later-generation phones) use SS.

Spread spectrum is primarily a digital communication system. SS is being used for some voice systems (including DECT cordless phones), but the analog voice is first sampled and digitized before transmission).

Recall from earlier discussions of modulation that it is possible to increase the S/N ratio at the output of certain types of demodulators if more channel bandwidth is used. This was certainly true for WBFM, and was also true on digital links using error correction. It therefore seems intuitive that an SS system that intentionally occupied a wide bandwidth might also offer some type of output S/N improvement.

In the late 1940's, Claude Shannon published his famous expression relating a communication channel's capacity to convey information as follows:

$$C = W \log_2\left(1 + \frac{S}{N}\right)$$

where C = channel capacity in error-free bits per second
W = channel bandwidth in Hz
N = noise power
S = signal power

This important equation shows the relationship between the ability of a channel to transfer error-free information compared to the channel bandwidth and the S/N ratio. For very small S/N ratios, as might be found in noisy communication channels, the expression can be simplified, approximated, and re-arranged to yield:

$$W = \frac{NC}{S}.$$

This indicates that for any given data rate, we can decrease the required S/N ratio by increasing the bandwidth used to transfer the information.

There are two types of SS systems: frequency hopping (FHSS), and direct sequence (DSSS). A frequency hopper is easy to understand when you remember that one of the original goals was to allow communication in the presence of jamming. Imagine that a transmitter and a receiver each have frequency synthesizers which allow the transmit and receive frequencies to move to one of perhaps 100 discrete frequencies in the band. Imagine that both transmitter and receiver were microprocessor controlled, and that the clocks were stable and had somehow or other achieved synchronization. Software routines in both processors generated pseudo-random sequences which are used to cause the transmission frequency to appear to hop about in a random fashion amongst the 100 channels. Because transmitter and receiver are synchronized, they both hop to the next frequency at the same time, so there is no interruption in the data flow. Anyone listening with a conventional receiver would only hear occasional bursts of noise.

A direct sequence system does not use a quantity of discrete frequencies. Instead, a single carrier is intentionally spread by digitally modulating (usually bi-phase modulation) the carrier with a high speed pseudo-random sequence. The net effect is that the many sidebands are spread out over a wide bandwidth (often several MHz). The total transmitter power is spread over this bandwidth, so anyone listening with a conventional receiver would just hear an overall increase in background noise. The data is added to the bit stream by selectively inverting (using an exclusive or gate) the polarity of the pseudo-random pattern. At the DSSS receiver, the same pseudo-random sequence is used to **correlate** with the incoming signal: assuming that the sequences are synchronized, it is possible to recover the original data. A block diagram of a DSSS system is shown below:

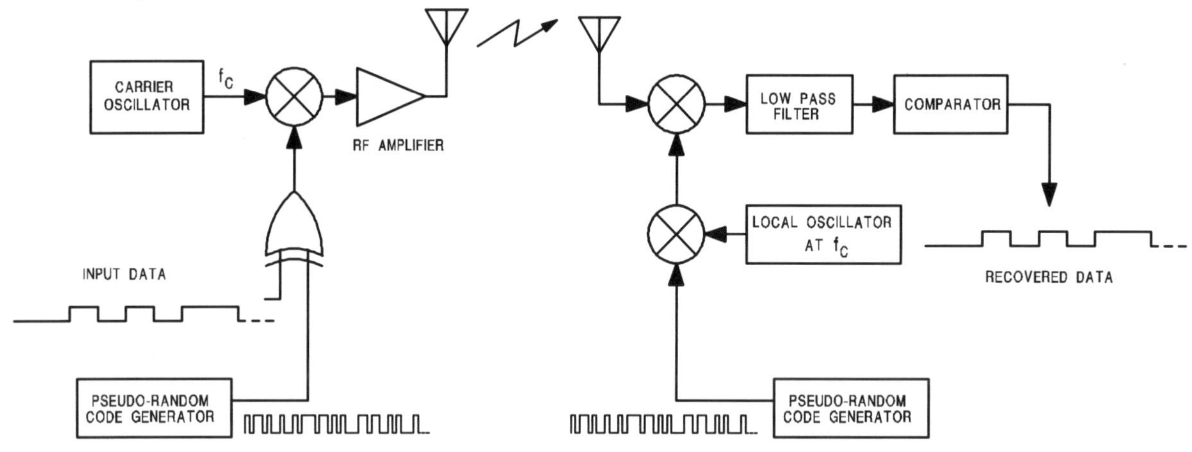

The transmitted spectra of both FHSS and DSSS systems are shown below:

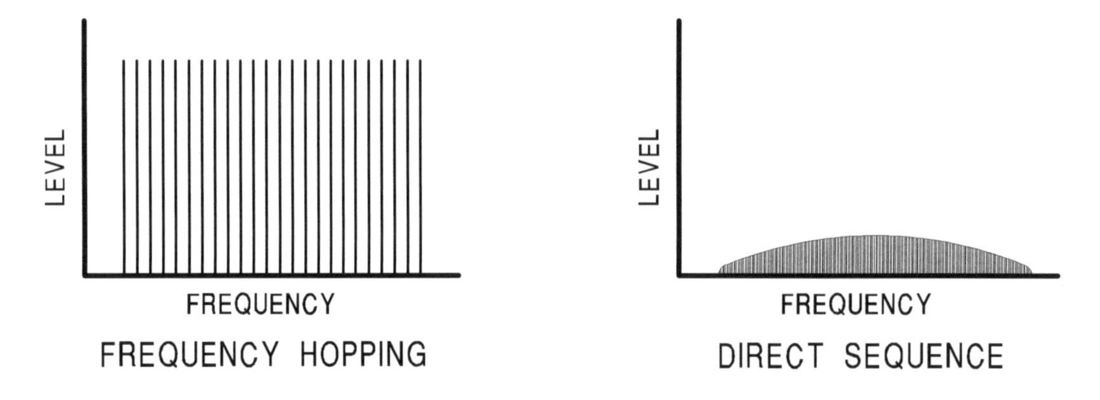

Narrow band users of the communication channel will be largely unaffected by the presence of an SS signal. Conversely, SS users can continue to communicate, even though there may be narrow band systems also transmitting. If different pseudo-random sequences are used, it is possible to have several SS systems sharing the same set of frequencies.

The improvement of S/N ratio offered by a specific SS system is called **process gain**. It can be estimated simply by dividing the RF bandwidth consumed by the data rate (bps) conveyed.

In North America, SS can be used **without a user license** to transmit digital data, so long as the equipment is type-accepted. There are many manufacturers of this sort of **wireless LAN** equipment: some are hoppers, others are direct sequence. The frequency bands permitted are 902 - 928 MHz , 2.4 GHz and 5.7 GHz. The various Wi-Fi standards that are in common use are based on SS, and the standards specify not only frequencies and modulation type, but also data throughput and data protocol.

ANTENNA GAIN, DIRECTIVITY, AND PATH LOSS

In the amateur radio community, the topic of antennas is subject to all kinds of erroneous information, folklore, witchcraft, and "old wive's tales". Because hams seldom have the luxury of having large amounts of real estate with which to do true comparison testing, the relative merits of one design vs another are typically established by here-say.

Probably the most authoritative text book on antennas was written by Dr. John D. Kraus: it is entitled "Antennas". Much of the information to follow has been extracted from this comprehensive text.

An antenna is a transition device between a *guided wave*, as might exist in a transmission line, and a *free-space wave*. An antenna appears as an impedance to the transmission line, having a resistive component R_r called the **radiation resistance**. The radiation resistance is not an actual resistance of part of the antenna, but is due primarily to the physical antenna in its environment.

The radiation pattern of an antenna is different at close distances than it is at long distances. At distances that are large compared to the size of the antenna, and large compared to a wavelength, the shape of the radiation pattern is independent of distance: this is the so-called **far-field** condition. The boundary between the near-field zone and the far-field zone is approximately defined as:

$$R = \frac{2L^2}{\lambda}$$

where R = the distance (radius) from the antenna in metres
L = the maximum dimension of the antenna in metres
λ = wavelength in metres

In the far field, the received intensity falls off as the square of the distance. In the near field, the rate of fall off is higher.

A theoretical transmitting antenna that has no physical size, yet radiates the energy uniformly in all directions is called an **isotropic** antenna. Imagine an isotropic antenna being located in the centre of a sphere. The surface of the sphere will be uniformly illuminated by power from the antenna: the amount of power per unit area anywhere on the sphere's surface will be identical. At a given distance from the isotropic radiator, if we wish to intercept more of the radiated power, we must collect from a larger area on the sphere. In order to be able to make this kind of calculation independent of the actual distance from the antenna, we deal in **solid angles**.

If you consider that the antenna is at the centre of an apple, the solid angle is a measure of the angular size of the three-dimensional wedge that could be cut out with a knife. Imagine drawing a circle on the surface of the apple, with area A. If you carefully cut a cone from the centre of thc apple to the outside of the line that you had drawn on the surface, the solid angle of the cone would be A/r^2, where r is the radius of the apple. As you think about it, the shape of a cone of a given solid angle will be independent of the diameter of the apple. Solid angle is measured in **steradians**.

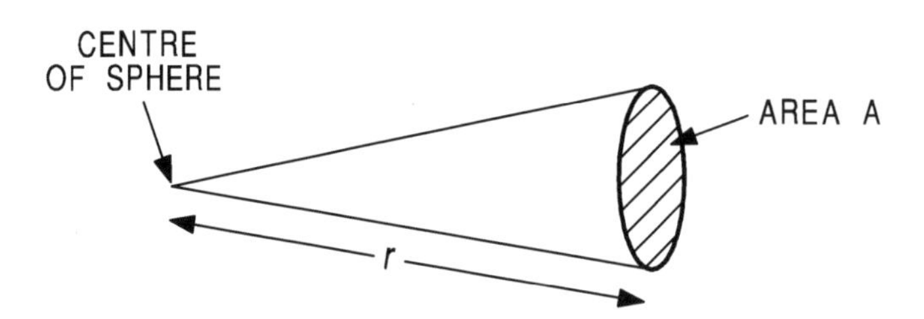

A sphere has a surface area of $\mathbf{4\pi r^2}$. A sphere has a total solid angle of $\mathbf{4\pi}$ steradians. If an isotropic antenna is radiating power, the fractional power intercepted by an area A at a distance r is simply equal to the ratio of the solid angles, or $\frac{A}{4\pi r^2}$.

In the real world, there is no such thing as an isotropic antenna. Even a so-called omnidirectional antenna will radiate more power in one direction or another. It should therefore be intuitive that in certain positions, a given antenna will radiate more power per steradian than in other positions. This gives rise to the concepts of **directivity** and **gain**. If it was possible to make an isotropic antenna, consider what would happen if it were mounted just above an infinite ground plane: all the power would have to be radiated in a hemisphere of 2π steradians. That therefore means that a given solid angle will now "contain" twice as much power as that which existed in the case where there was no ground plane. We can therefore say that the gain of this antenna is 3 dB_i, which means that at the centre of its main lobe, the power density is 3 dB higher than it would be from an isotropic antenna in free space. The beamwidth would obviously be 180 degrees.

The directivity D of an antenna is the ratio of the maximum radiation density (measured in Watts per steradian) to the average radiation intensity averaged over a sphere surrounding the antenna. All measurements are to be made in the far field. The beamwidth of an antenna's radiation pattern is often expressed in terms of degrees in the horizontal and vertical directions (to the 3 dB points). Ignoring minor lobes, the directivity can be approximated as follows:

$$D \approx \frac{41{,}000}{HBW \cdot VBW}$$

where: D = the directivity (a dimensionless number)
HBW = the horizontal 3 dB beamwidth in degrees
VBW = the vertical 3 dB beamwidth in degrees

Directivity and gain are interrelated: $G = kD$, where k is the efficiency factor of the antenna, relating to ohmic losses in the antenna. The value of k is between 0 and 1.

It can therefore be seen that the gain of an antenna is inversely proportional to the beamwidth.

Antennas are **reciprocal** devices. The observed gain when used as a receiving device is the same as the gain when used as a transmitting antenna. It is sometimes hard to understand how this can be possible unless you also understand the concept of **effective aperture**.

Imagine for a moment that an isotropic antenna is connected to a transmitter of power W. Any sphere that is drawn around this antenna will have a uniform power density of $W/4\pi$ Watts per steradian. At a given distance from the source, the amount of power which can be collected is proportional to the area of the collector. It is obvious that at a given distance from the radiating antenna, the only way to increase the amount of power intercepted by a receiving antenna is to increase the area over which the power is collected. A receiving antenna must intercept power over an area: the amount of area is called the **effective aperture**.

For a simple antenna like a half-wave dipole, it is easy to imagine that the effective aperture is defined by a rounded rectangle or an ellipse which is probably about as long as the antenna, and has a smaller dimension in the other axis. This is indeed the case, as illustrated by the following diagram:

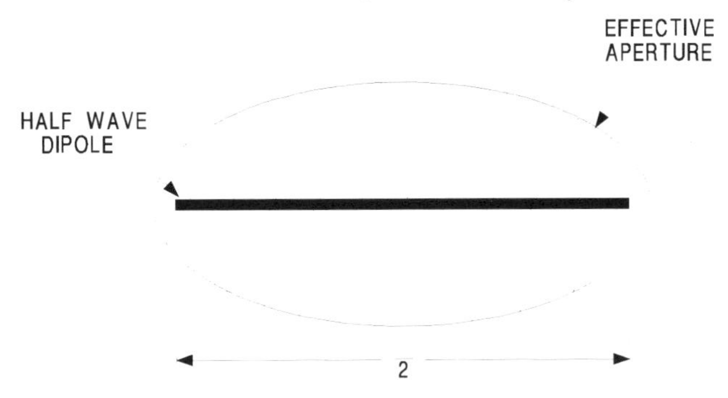

In free space, the area of the effective aperture of a half-wave dipole is approximately $0.13\lambda^2$. This "feels" all right, but note that the effective aperture is smaller with higher frequencies.

In order to provide gain over a dipole, a receiving antenna must have an effective aperture which is larger than $0.13\lambda^2$. A yagi with 6 dBd gain must have an effective aperture of four times that of a dipole, or $0.52\lambda^2$. Note however, that when looking toward a yagi (end-on to the boom), the physical area doesn't look much different from a dipole. The difference is that it is intercepting energy from a larger surrounding area than a dipole would.

Since antennas are reciprocal devices, it is clear that gain and effective aperture must be somehow related. The relationship is as follows:

$$G = \frac{4\pi}{\lambda^2} A_e$$

where G = antenna gain
A_e = effective aperture

Path Loss

We now have enough information to calculate how much power is present at a receiving antenna, assuming that we know the transmitted power, the distance between the transmitting and receiving antennas, and the antenna gain on both ends. It is simply a matter of geometry and mathematics. In 1946, Harald Friis of Bell labs developed the **Friis Transmission Formula**, which is shown below:

$$P_r = P_t \frac{A_{et} A_{er}}{r^2 \lambda^2}$$

where: P_r = the received power in Watts
P_t = the transmitted power in Watts
A_{et} = the effective aperture of the transmitting antenna in square metres
A_{er} = the effective aperture of the receiving antenna in square metres
r = distance between antennas in metres
λ = wavelength in metres

The above important equation assumes that the antennas are matched, and that the measurements are made in the far-field zone. This equation can be re-written as a **path loss** from the transmitting antenna's terminals to the output of the receiving antenna:

$$\text{Loss(dB)} = 32.45 + 20\log(d) + 20\log(f) - G_T - G_R$$

where: d = distance between antennas in kilometres
f = frequency in MHz
G_T = gain of transmitting antenna in dB
G_R = gain of receiving antenna in dB

This is a very interesting equation. In particular, note that the total path loss increases with frequency. If the frequency is increased 10 times, the path loss increases by 20 dB! Based on this, one would assume that the signals get weaker and weaker as you move from 2 metres to 70 cm and on up. If this is the case, how is it that you work EME using 100 Watts on 2304 MHz, whereas most 2 metre moonbouncers need a kW? This is partly due to the fact that as the frequency increases, antennas get physically smaller for the same amount of gain. If the antenna's physical size is kept constant as the frequency is increased, the resulting larger value of gain will offset the increased path loss.

> An extremely rough rule of thumb (only partly tongue-in-cheek) is: *If the antenna's physical bulk (as measured by its weight in kg) is kept constant, the path loss is approximately independent of frequency!* This is because at higher frequencies, the wavelength is smaller, and antennas of a given gain would become physically smaller. Therefore, at higher frequencies, a given physical "bulk" of antenna will give more gain (and directivity), thereby offsetting the increased path loss. Note that this ignores the effect of any frequency-dependent atmospheric absorption.

Even though you can keep the overall received signal strength constant as you increase frequency simply by putting up more aluminum in the sky, another issue needs to be carefully considered: directivity. As you use higher and higher gains at the UHF and microwave frequencies, the antenna's beamwidth gets so narrow that it is often hard to point the antennas effectively! At 10 GHz, the major problem in "mountain-topping" is aligning the two antennas: accurate compasses, theodelites, and telescopes are the order of the day.

Above 2 GHz, dish antennas become quite common. A dish is simply a section of a parabaloid, with a small antenna at the focus. Received signals that hit the dish are reflected back to the focus and collected.

This is very similar to the way a reflector in a flashlight works. The diameter of the dish is much larger than a wavelength, so optical ray tracing techniques can be used in designing the reflector. With this type of antenna, the effective aperture is equal to the diameter of the dish: this makes it easy to understand the interrelationship between gain and effective aperture.

Ignoring dish antennas for the moment, let's discuss what allows an assembly of aluminum tubes to be able to produce gain relative to an isotropic antenna. Basically, any antenna which does not radiate power perfectly uniformly in all directions must have directivity, and hence must have "gain". A half wave dipole antenna in free space has a characteristic "doughnut-shaped" radiation pattern: at a point in the far field that is perpendicular to the antenna, the power density is higher than it would have been from an isotropic antenna by 2.15 dB. It is sometimes convenient to express the relative gains of antennas by comparing them to a half wave dipole: in that case, the gain should be stated in terms of dB_d instead of dB_i.

Consider two half wave dipoles that are mounted parallel to each other, but separated by a quarter wavelength. We will connect a signal source each of the two dipoles via two transmission lines. The transmission line to the left dipole is one quarter wavelength shorter than the line to the right hand dipole, after accounting for velocity factor. It can be seen that signals going from left to right will experience constructive interference at the right hand dipole, whereas signals going from right to left will suffer destructive interference at the left hand dipole. We will therefore have a radiation pattern that is predominantly toward the right. This structure will give gain over a dipole. The direction of radiation can be reversed by interchanging the feedlines, or by some other means that can change the relative phase of the signals at the two antennas. For something as simple as this, mechanical relays could be used to "switch in" a length of transmission line that has the correct electrical length. Military **phased array** radar antennas consist of several hundreds of fixed mount antennas, each connected to an electrically controllable phase shifter: The effective beam direction can be changed almost instantaneously in either the vertical or the horizontal planes.

Rather than connect multiple dipoles with carefully adjusted phasing lines, the yagi antenna uses a simpler approach. A single dipole is used, but the other elements are not electrically connected; indeed, they are referred to as **parasitic elements**. The parasitic elements are in the near field of the driven element, and current is coupled into them. The lengths of the parasitic elements is adjusted to give the desired phase relationship, and constructive interference occurs in the desired direction.

Government regulations are beginning to surface regarding the maximum **field strength** that can be allowed on adjacent properties. Radio waves are electromagnetic waves, and they can be specified either by the their electrical or magnetic fields. The relationship between the two fields is set by the impedance of free space, which is approximately 377 ohms. It is usual to quote electrical field strength in terms of volts per metre.

It is extremely complicated to calculate the field strength produced by a multi-element antenna in an arbitrary direction. It is fairly easy however, to calculate the electrical field strength of a half wave dipole in a perpendicular direction. The following approximation assumes a resonant antenna in free space:

$$E \approx \frac{7\sqrt{P}}{d}$$

where E = field strength in volts per metre
P = power delivered to the antenna in Watts
d = distance form the antenna in metres

Note that this approximation assumes the antenna is located in free space. Ground effects and localized reflections can easily cause variations exceeding 2 in the actual field strengths.

SAFETY CONSIDERATIONS WITH RF FIELDS

At VHF frequencies and higher, the field strengths produced by amateur transmitters can cause some safety concerns. It would certainly not be prudent to stand directly in front of a high gain antenna being fed 10 Watts or more at 2400 MHz! At higher frequencies, the primary consideration is absorption of energy by the body, and the consequent heating effects. At lower frequencies, high electric field strengths can stimulate the body's nerves.

This is a very controversial topic, subject to much debate and some litigation. The following information is extracted from "Mobile Antenna Systems Handbook", by Fujimoto and James.

Because of the long wavelengths used, the HF frequencies are not a major concern, even if full amateur legal power is used. Low band transmitting antennas are typically mounted dozens of metres away from personnel, and heating effects would be minuscule.

The major concern for amateur radio operators relates to the use of hand held transceivers on frequencies between 144 and 1296 MHz. The mode of usage dictates that the antenna is often right next to the operator's head. If operating from a 12 volt source, some of these radios are capable of producing almost 10 Watts of RF output.

As of 1991, the ANSI C95.1-1982 standard was in the process of being revised. It was proposed to set two limits of human **absorption** in the 30 MHz to 15 GHz range:

1. *Controlled Exposure Limits*: 0.4 mW/g averaged over the whole body, or 8 mW/g averaged over a single gram of tissue. The averaging time is 6 minutes.

2. *Uncontrolled Exposure Limits*: 0.08 mW/g averaged over the whole body, or 1.6 mW/g averaged over a single gram of tissue. The averaging time is 30 minutes.

The term "Controlled Exposure" refers to the fact that the operator is aware of the fact that he is using a device that produces RF energy, and has the ability to turn it off. The "Uncontrolled Exposure Limits" are intended to cover the case where individuals have no knowledge of the presence of the RF field.

To determine how much power might be absorbed by an operator using a hand held VHF or UHF radio held (as it usually is) close to the head, a fully instrumented dummy was carefully measured under controlled conditions. Some of the interesting data was reported as follows:

- Using a 6.4 Watt radio with a rubber duck antenna at 150 MHz held 5 cm from the lips, a peak exposure of 0.5 mW/g was observed just above the eyebrow. The average brain exposure was 0.1 mW/g.

- Using a 6.4 Watt radio with a quarter wave antenna at 450 MHz held 5 cm from the lips, a peak exposure of 1.2 mW/g was observed in the eyeball. The average brain exposure was 0.15 mW/g.

When the test was repeated using a "rubber duck" antenna, the eyeball exposure dropped to 0.9 mW/g, and the average brain exposure was 0.12 mW/g.

- Using a 6.4 Watt radio with a sleeve dipole antenna at 900 MHz held 5 cm from the lips, a peak exposure of 0.7 mW/g was observed at the skull-brain interface. If the antenna is held only 2 cm from the lips, a peak exposure of 3 mW/g was observed in the eye.

There is much debate as to what this data really means, or how big the risk factors are. However, it would seem prudent when using hand held VHF or UHF transceivers to use a remote speaker/microphone to minimize exposure of the head to RF fields! If the antenna is several feet away from critical parts of the body, the exposure risk should be very small. In mobile installations using external antennas, the vehicle's metal body will provide a large degree of shielding to the occupants, especially if the antenna is mounted on the centre of the roof or trunk.

Power Density Limits

The above-referenced ANSI standard was the basis for future work which has now become IEEE C95.1–2005, also referred to as the ***human exposure standard***. The complete name is "Standard for Safety Levels with Respect to Human Exposure to Radio Frequency Electromagnetic Fields, 3 kHz to 300 GHz". This standard sets specific permitted limits for electric and magnetic field strengths (measured as Volts per metre, and Amps per metre respectively), as well as limits on RF **power density** (measured in Watts per square metre). Maximum Permitted Levels of RF power density in this **IEEE** standard are:

Controlled Environments

Frequency Range (MHz)	Maximum Permitted Power Density (Watts/metre^2)
1-30	$9{,}000/f^2$
30-300	10
300-3,000	f/30

Uncontrolled Environments

Frequency Range (MHz)	Maximum Permitted Power Density (Watts/metre^2)
1.34-30	$1{,}800/f^2$
30-400	2
400-2,000	f/200

In 2015, **Health Canada** published their "Safety Code 6" which also set limits on maximum permitted field strength (both electric and magnetic) and power density. The power density limits specified by this standard are as follows:

Controlled Environments

Frequency Range (MHz)	Maximum Permitted Power Density (Watts/metre^2)
10-20	10
20-48	$44.72/f^{0.5}$
48-100	6.455
100-6000	$0.6455\ f^{0.5}$

Uncontrolled Environments

Frequency Range (MHz)	Maximum Permitted Power Density (Watts/metre^2)
10-20	2
20-48	$8.944/f^{0.5}$
48-300	1.291
300-6000	$0.02619\ f^{0.6834}$

In 1997, the **FCC** issued its own set of limits on RF power density as follows:

Controlled Environments

Frequency Range (MHz)	Maximum Permitted Power Density (Watts/metre^2)
1.34-30	$900/f^{2}$
30-300	1
300-1,500	f/300

Uncontrolled Environments

Frequency Range (MHz)	Maximum Permitted Power Density (Watts/metre^2)
1.34-30	$180/f^{2}$
30-300	0.22
300-1,500	f/1,500

It should be noted that the different organizations referenced above (Health Canada, IEEE, FCC) all set different limits! These are summarized in the graphs below. Note that both axes use logarithmic scales.

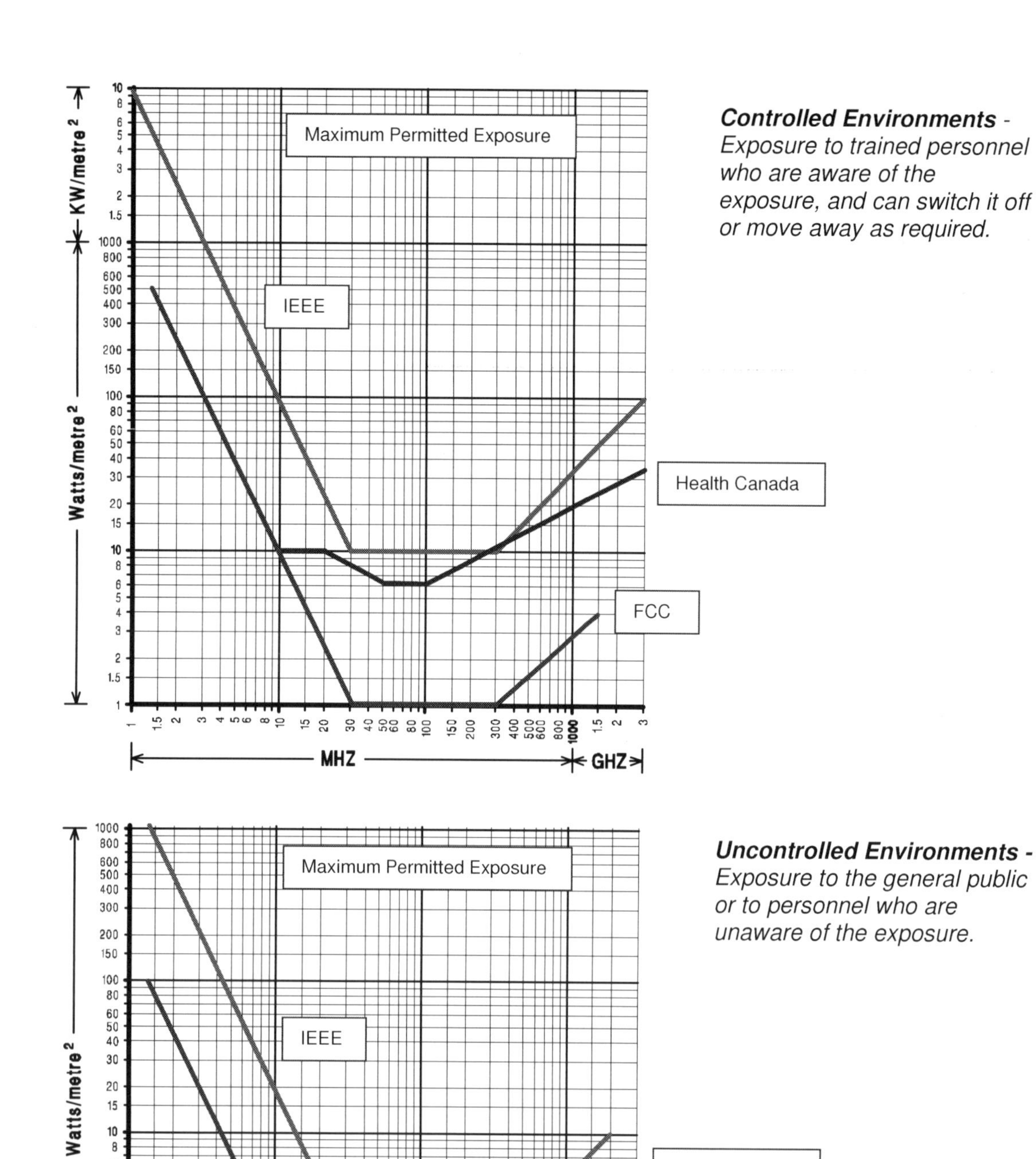

Controlled Environments *- Exposure to trained personnel who are aware of the exposure, and can switch it off or move away as required.*

Uncontrolled Environments *- Exposure to the general public or to personnel who are unaware of the exposure.*

RF SMALL SIGNAL AMPLIFIERS

The object of a small signal RF amplifier is to increase the magnitude of low-level signals without adding appreciable noise or distortion products. Small signal amplifiers are operated in "Class A", which means that the active device (tube, FET, or BJT) is always conducting during the cycle of the input signal - although the current through the device will vary, it will never go to zero. Efficiency (RF power output versus DC power input) is not usually a concern, and output power levels rarely exceed 0 dBm.

Small signal amplifiers are usually designed using either bipolar transistors or FETs. We already discussed how to set up the bias networks for these devices; we will now concentrate primarily on RF considerations. Because impedance levels differ considerably at different points in an amplifier, we will mostly concentrate on **power gain** rather than voltage gain as we analyze circuits.

At RF frequencies, a model of a bipolar transistor must include the effects of lead inductance and feedback capacitance (between collector and base). As the frequency changes, both the input and the output impedance change. Because of the feedback capacitance, changes of load impedance will effect the input impedance, and vice-versa. Certain combinations of load and source impedance may combine with the feedback capacitance to cause an unstable condition at some frequency, and the amplifier can then become an oscillator.

As the frequency of an input signal is increased from a low value, an amplifier's power gain starts to roll off at 6 dB per octave, until it falls to 0 dB at a frequency known as $\mathbf{f_{max}}$. The gain provided in a transistor RF amplifier is reduced by the feedback capacitance: if external components are used to cancel out the internal feedback, the resulting gain is referred to as the **neutralized power gain**.

Manufacturer's data sheets provide information on a transistor's input and output impedance at various frequencies for specific circuit configurations (common emitter, common base, or sometimes common collector). Sometimes Y parameters (admittances) are used, but the most popular and convenient method is to use S-parameters. As discussed in an earlier chapter, there are four S parameters, and they represent the input and output reflection coefficients, and the forward and reverse transmission coefficient, all referred to a 50 ohm measuring impedance. Measurements are made using 50 ohm source and load impedances: both magnitude and phase are specified. In order to achieve maximum gain from a transistor amplifier, the input and output impedances must be matched to their complex conjugates.

* *At this point, you may wish to consider how deeply you want to get into the mathematics involved in designing high performance RF amplifiers. The following section will involve lots of operations with complex numbers, and the reader may wish to skip ahead if this is too much detail. Commercial software is available that can easily do much of the required analysis, or the application might be implemented using some of the 50 ohm pre-matched amplifier gain blocks that are available from a variety of sources. If you still want to know how to design small signal amplifiers using classical manual techniques, grab a pencil and push on!*

An example of a set of S parameters provided on a manufacturer's data sheet is shown below. This example is a UHF transistor (the NE85632 in TO-92 package) used in the common emitter configuration at a V_{CE} of 10 V, and a collector current of 10 mA:

Frequency	S_{11}		S_{21}		S_{12}		S_{22}		K	MAG
(MHz)	MAG	Angle	MAG	Angle	MAG	Angle	MAG	Angle		(dB)
100	0.62	-58	20.35	135	0.02	70	0.8	-26	0.31	30.1
200	0.45	-95	13.62	113	0.03	55	0.59	-36	0.71	26.6
500	0.35	-141	6.44	89	0.07	63	0.39	-42	0.93	19.6
1000	0.31	-177	3.46	65	0.13	60	0.36	-51	0.99	14.3
1500	0.31	160	2.46	48	0.19	57	0.35	-70	0.98	11.1
2000	0.34	138	2.04	30	0.25	46	0.35	-88	0.91	9.1

Note that for this transistor the manufacturer has already calculated the Rollett Stability Factor K for us. Recall that K is defined as:

$$K = \frac{1 + |D_S|^2 - |S_{11}|^2 - |S_{22}|^2}{2 \cdot |S_{21}| \cdot |S_{12}|}, \quad \text{where } D_S = S_{11}S_{22} - S_{12}S_{21}$$

If K is less than one, the transistor is *potentially* unstable at certain combinations of source and load impedance. If K is greater than one, the transistor is unconditionally stable. Looking at this data, it can be seen that you would have to be careful when designing an amplifier using this transistor with tuned input and output circuits: there is the possibility of creating an oscillator at certain settings of the capacitors! You can either proceed carefully, choose another set of bias conditions that yield different S parameters, choose another package configuration, or choose another, "tamer" transistor.

If it was desired to build an amplifier that provided at least 12 dB of gain at 1296 MHz and that was unconditionally stable, a reasonable choice would be the NE85635, which is actually the same semiconductor die as the previously-described device, but instead uses a Micro-X package. When biased at 10 volts and 20 ma, the characteristics are as follows:

Frequency	S_{11}		S_{21}		S_{12}		S_{22}		K	MAG
(MHz)	MAG	Angle	MAG	Angle	MAG	Angle	MAG	Angle		(dB)
100	0.65	-78	34.2	141	0.003	9	0.79	-27	1.24	37.6
500	0.64	-162	10.38	93	0.007	43	0.26	-49	3.66	23.1
1000	0.62	179	5.32	75	0.02	52	0.21	-56	2.69	17.1
1500	0.62	167	3.62	63	0.04	53	0.23	-56	1.94	14.0
2000	0.62	153	2.8	50	0.06	54	0.19	-63	1.74	11.7
2500	0.64	140	2.22	39	0.11	50	0.25	-82	1.12	10.9
3000	0.65	131	1.92	26	0.14	48	0.27	-94	0.99	11.4
3500	0.67	120	1.62	14	0.16	38	0.28	-105	0.96	10.1
4000	0.68	110	1.48	4	0.17	38	0.3	-115	0.96	9.4

The column labeled "MAG" gives manufacturer's values for "Maximum Available Gain", which is defined as $MAG = \frac{|S_{21}|}{|S_{12}|}\left(K \pm \sqrt{K^2 - 1}\right)$. In practice it is impossible to achieve gains that match MAG, but with careful design it is possible to come reasonably close: use MAG as a guideline in selecting a suitable transistor for an application, but expect that the actual achieved gain will be a couple of dB lower. For this example application, it can be seen that the transistor is stable over a wide frequency range surrounding the frequency of interest, and MAG seems more than sufficient. A plot of S_{11} and S_{22} as a function of frequency in GHz for this transistor is presented below:

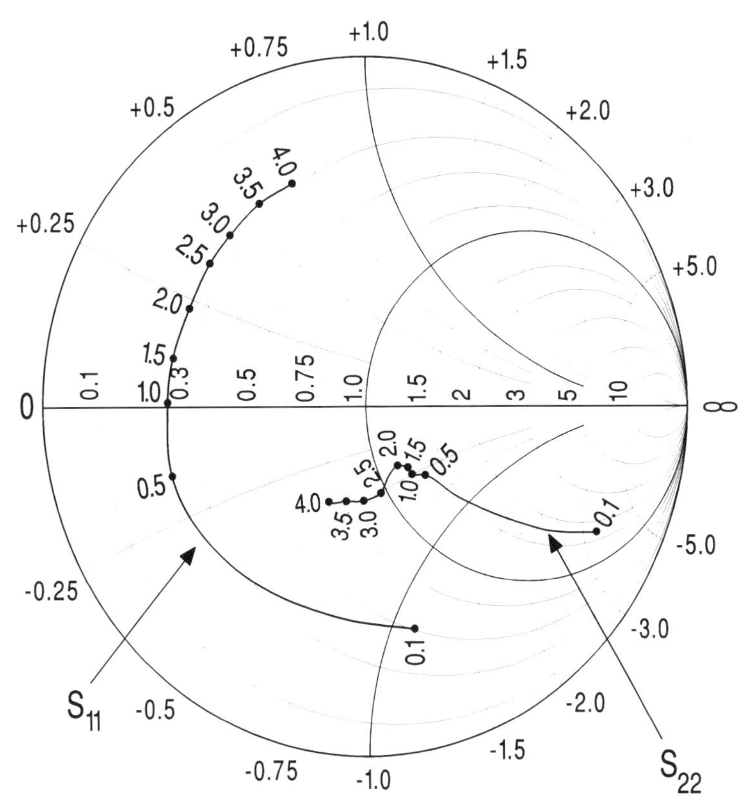

Once S_{11} and S_{22} (the input and output reflection coefficients, referred to 50 ohms) are plotted on a Smith Chart, the corresponding impedances or admittances can be directly read off of the appropriate scales. Note that S parameters for a transistor or FET vary with package type, frequency, bias voltage, and bias current. For this example, the terminated input and output impedances at our design frequency of 1296 MHz are as follows:

S_{11} = 0.62 at 170 degrees, or an input impedance of (12 + j5) Ω.

S_{22} = 0.22 at -56 degrees, or an output impedance of (54 - j21) Ω.

Note that both of these values assume that the opposite port is terminated in 50 ohms, not the proper conjugate impedance! The actual input and output impedance will NOT be those values described above, because S_{12} is not equal to zero (in other words, the output impedance affects the input impedance, and vice versa).

We need to determine the input and output reflection coefficients of this amplifier when it is conjugately matched. In other words, we want to know what Γ_{IN} is when the output is terminated in Γ^*_{OUT}, and we want to know what Γ_{OUT} is when the input is terminated in Γ^*_{IN}. It is possible to determine the input and output reflection coefficients of an amplifier when the terminating impedance of the other port is known by using the relationships:

$$\Gamma_{IN} = S_{11} + \frac{S_{12}S_{21}\Gamma_L}{1 - S_{22}\Gamma_L}$$

$$\Gamma_{OUT} = S_{22} + \frac{S_{12}S_{21}\Gamma_S}{1 - S_{11}\Gamma_S}$$

Where Γ_{IN} = input reflection coefficient of the amplifier
Γ_{OUT} = output reflection coefficient of the amplifier
Γ_S = reflection coefficient of the source
Γ_L = reflection coefficient of the load

The above relationships are useful, but they don't tell us what the actual conjugate impedances to use are. The following expressions define the actual source and load reflection coefficients to use in order to achieve a **simultaneous conjugate match** at the input and output of the transistor amplifier:

Define:

$$D_S = S_{11}S_{22} - S_{12}S_{21}$$
$$B_1 = 1 + |S_{11}|^2 - |S_{22}|^2 - |D_S|^2$$
$$B_2 = 1 + |S_{22}|^2 - |S_{11}|^2 - |D_S|^2$$
$$C_1 = S_{11} - D_S S^*_{22}$$
$$C_2 = S_{22} - D_S S^*_{11}$$

Then, for a simultaneous conjugate match -

$$\Gamma_S = \frac{B_1 \pm \sqrt{B_1^2 - 4|C_1|^2}}{2C_1}$$

$$\Gamma_L = \frac{B_2 \pm \sqrt{B_2^2 - 4|C_2|^2}}{2C_2}$$

A minus sign should be used in front of the radical if the coefficient B is positive, and vice versa.

We can use these relationships to calculate the required source and load reflection coefficients as seen by the amplifier. First we calculate the following intermediate values of D_S, B_1, C_1, C_2, and B_2 for the chosen operating frequency of 1296 MHz:

$\mathbf{D_S} = S_{11}S_{22} - S_{12}S_{21}$ $= (0.62 \angle 170)(0.22 \angle -56) - (0.03 \angle 52)(4.3 \angle 68)$
$= (0.1364 \angle 114) - (0.129 \angle 120)$
$= (-0.0555 + j0.1246) - (-0.0645 + j0.112)$
$= 0.009 + j0.0126$
$= \mathbf{0.0155 \angle 54}$

$$\mathbf{C_1} = S_{11} - (D_S\, S_{22}^{*}) = (0.62\angle 170) - (0.0155\angle 54)(0.22\angle 56)$$
$$= (0.62\angle 170) - (0.00341\angle 110)$$
$$= -0.611 + j0.108 + 0.00117 - j0.00320$$
$$= -0.610 + j0.105 = \mathbf{0.619\ \angle 170}$$

$$\mathbf{C_2} = S_{22} - (D_S S_{11}^{*}) = (0.22\angle -56) - (0.0155\angle 54)(0.62\angle -170)$$
$$= (0.22\angle -56) - (0.00961\angle -116)$$
$$= (0.123 - j0.182) - (-0.00421 - j0.00864)$$
$$= 0.127 - j0.173$$
$$= \mathbf{0.215\angle -54}$$

$$\mathbf{B_1} = 1 + |S_{11}|^2 - |S_{22}|^2 - |D_S|^2 = 1 + 0.384 - 0.0484 - 0.000240 = \mathbf{1.3354}$$

$$\mathbf{B_2} = 1 + |S_{22}|^2 - |S_{11}|^2 - |D_S|^2 = 1 + 0.0484 - 0.3844 - 0.000240 = \mathbf{0.664}$$

* Note that whenever an asterisk is present after a value, it means to take the complex conjugate of that value. Now we can use these values to calculate the required load reflection coefficient Γ_L and source reflection coefficient Γ_S using the following expressions:

$$\Gamma_S = \frac{B_1 \pm \sqrt{B_1^2 - 4|C_1|^2}}{2C_1}$$

$$= \frac{1.3354 - \sqrt{(1.3354)^2 - 4(0.619)^2}}{2(0.619\angle 170)}$$

Note that the negative sign was used in front of the radical because B_1 is positive.

$$= \frac{0.835}{1.24\angle 170}$$

$$= 0.674\angle -170$$

and

$$\Gamma_L = \frac{B_2 \pm \sqrt{B_2^2 - 4|C_2|^2}}{2C_2}$$

$$= \frac{0.664 - \sqrt{(0.664)^2 - 4(0.215)^2}}{2(0.215\angle -54)}$$

Note that the negative sign was used in front of the radical because B_2 is positive.

$$= \frac{0.158}{0.43\angle -54}$$

$$= 0.367\angle 54$$

Note that the actual input and output reflection coefficients of the amplifier are the conjugates of Γ_S and Γ_L - this is the necessary condition for maximum power transfer and gain. This is shown on the following Smith Chart:

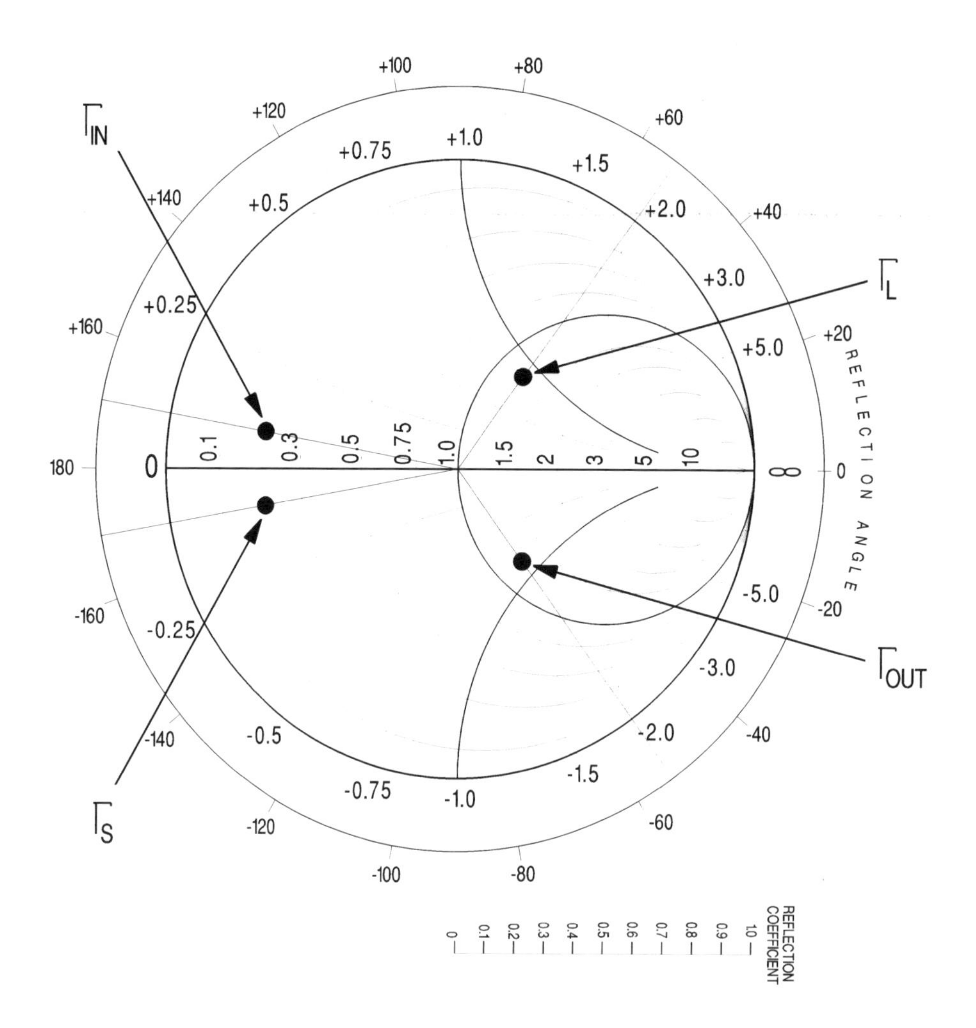

Phew! Once we have slogged through all of the above procedure in order to determine the required source and load reflection coefficients, we can then design an input and output matching network that will present the input and output of the transistor with the proper values of Γ_S and Γ_L. The matching networks can use either lumped components, transmission lines, stubs, or all three. We will now discuss the three different techniques.

Lumped Component Matching

First we will look at the use of a lumped component solution, using the methods discussed earlier:

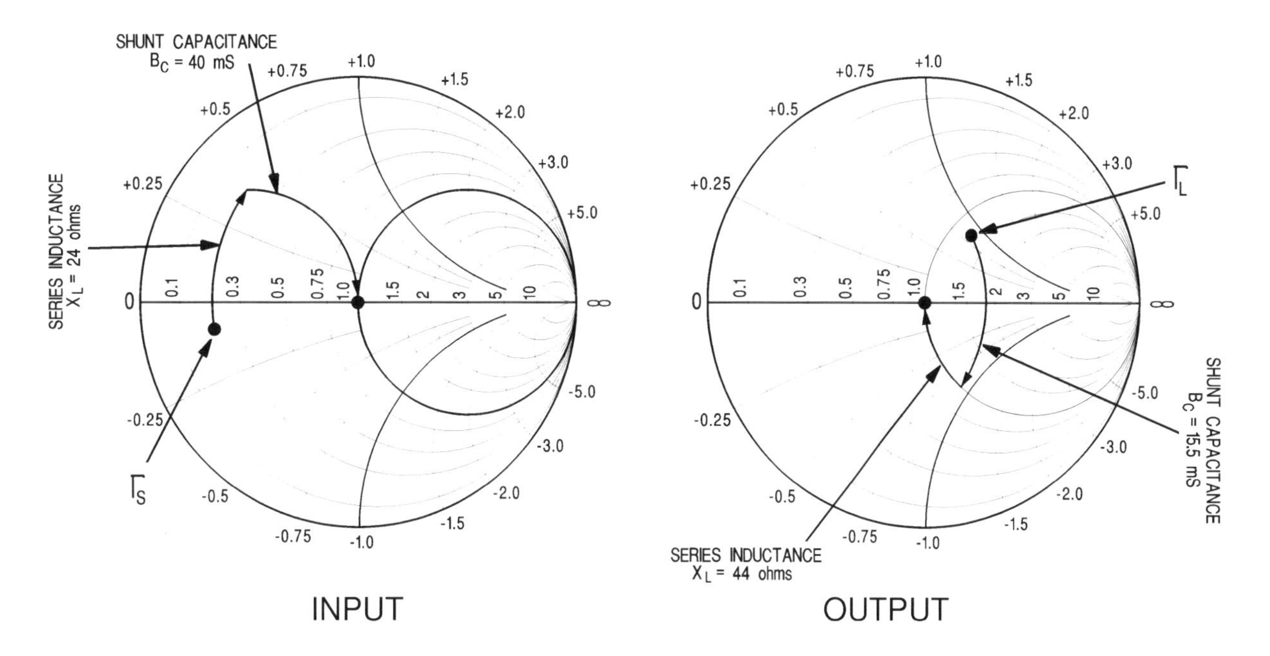

The transistor and its matching networks would look like the following circuit:

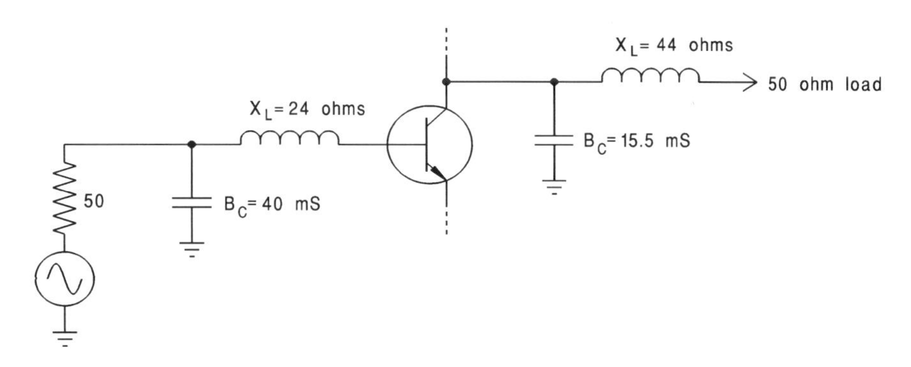

Note that biasing information is not shown in this schematic. The actual values of the matching components need to be chosen to give the required reactances and susceptances at the operating frequency of 1296 MHz. The values will be quite small, and this amplifier is probably best constructed using microstrip components using one of the following two approaches.

Transmission Line Matching

A matching solution using transmission lines can easily be worked out on a Smith Chart. The approach we will use consists of a short length of 50 ohm microstrip to bring the impedance back to the real axis, then a quarter wave transformer to get the impedance to 50 ohms:

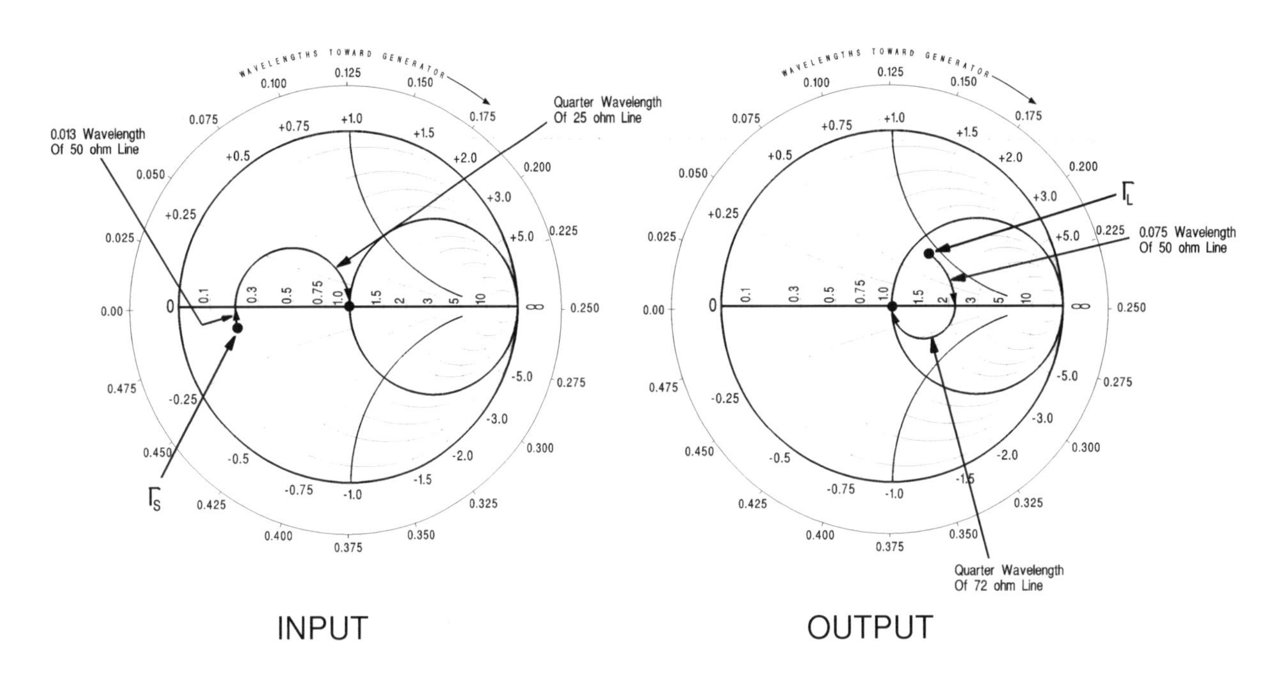

When implemented on a PCB artwork, the RF matching components will look like this:

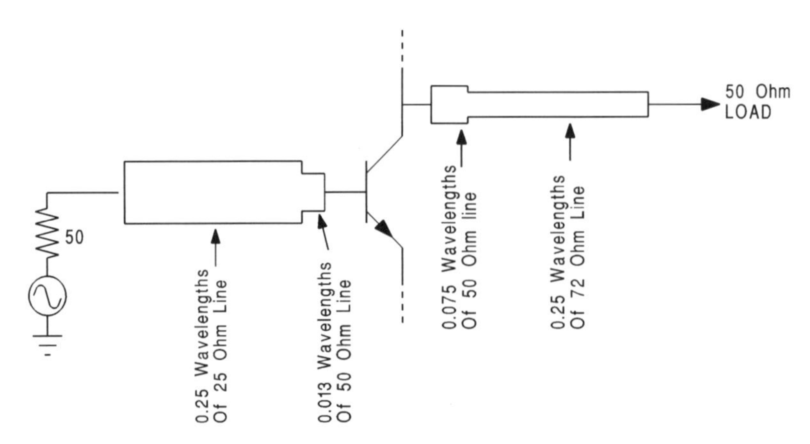

Note that the above diagram shows the top surface of the printed circuit board (where the components are mounted). It is assumed that the other side of the PCB is composed of un-etched copper, forming a continuous ground plane.

Stub Matching

Another microstrip approach uses sections and stubs of 50 ohm transmission line to perform the necessary matching. The first step is to convert the impedances represented by the values of Γ_S and Γ_L to admitance values Y_S and Y_L. The Smith Chart will then be used to plot admittances. On a constant SWR circle, rotate CCW to intercept the 1+j circle, then determine the length of an open-ended stub to nullify the remaining imaginary component:

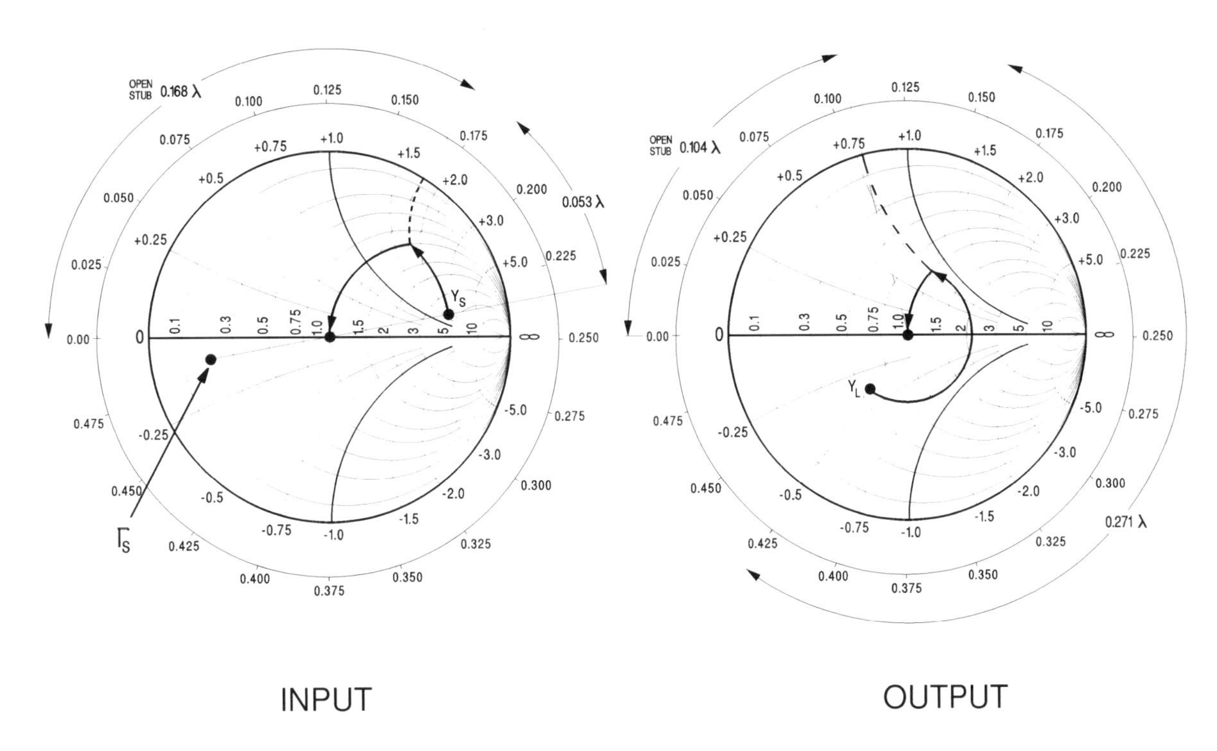

Using the data from the Smith Charts, the circuit layout can be implemented as follows:

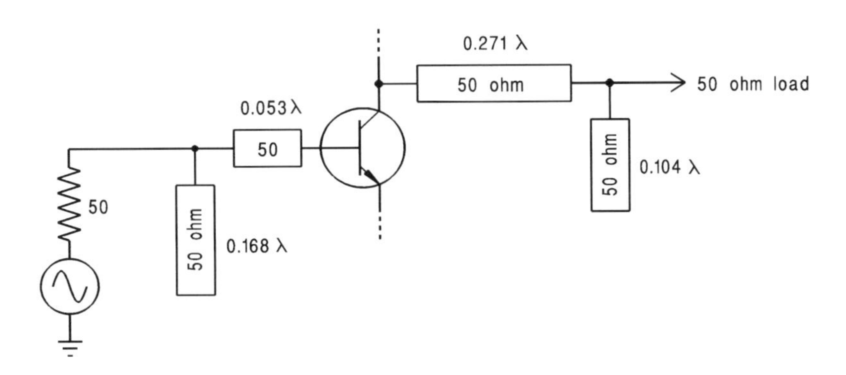

As can be seen by the above three examples, there are many different ways to design matching networks for transistor RF amplifiers, but there is a fair bit of work involved. Fortunately there is commercial RF design software (some of it is free) that can be of great assistance in this work!

DC Biasing

Note that the above discussion concerns only the RF matching circuitry. The designer still needs to set up DC circuitry to bias the transistor at the correct point. For frequencies of less than 50 MHz, the same biassing circuitry discussed early in this book can be used, although care must be taken to use extremely short leads on the emitter bypass capacitor. All bypass capacitors must be intended for RF use, with very low internal series inductance.

At higher frequencies it is almost always essential to directly ground the emitter lead of transistor amplifiers. An extremely short connection must be made to a solid ground point. A DC biasing arrangement is then required that will not adversely effect the RF performance, will allow the emitter to be grounded, and will hold a stable DC operating voltage and current. One approach (as applied to the stub-matched amplifier) is as follows:

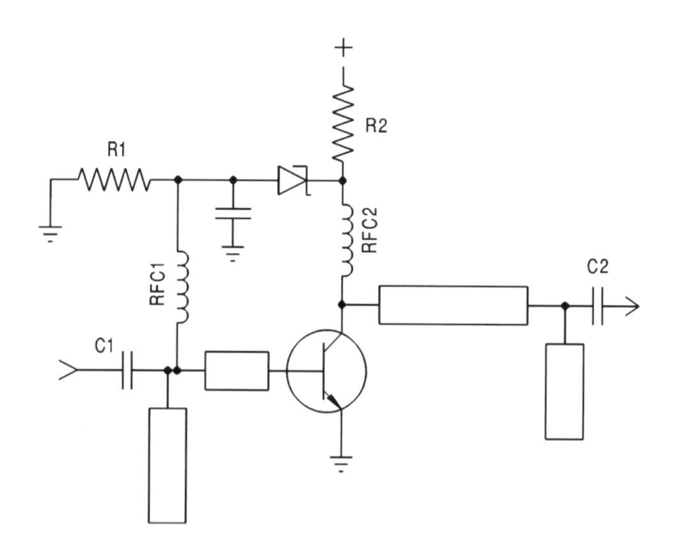

In this biasing arrangement, the two RF chokes RFC1 and RFC2 are inductors with a high reactance at the frequency of interest. For UHF and higher frequencies, they can often be made by using a fine trace of conductor on the PC board. The zener voltage is chosen to be approximately equal to the desired operating voltage across the transistor, and R2 is selected to give the required operating current, based on the supply voltage. From an RF standpoint, the biasing circuitry doesn't exist (because of the isolation from the RF chokes), but from a DC standpoint, there is very heavy negative feedback which will provide excellent stability.

The RF coupling capacitors C1 and C2 must have a low series reactance at the frequency of interest. Leads must be extremely short - a surface mount capacitor style is preferred. With a slight loss of DC stability, the zener diode can be replaced with a carefully chosen resistor.

Amplifiers are often used at the front end of a receiver, where the signal levels are extremely small. In this case, you are very concerned that the internal noise contributions of the active device not swamp out or degrade the input signal - in other words, you would like your amplifier to be a **low noise** design.

FETS and transistors are available that are meant specifically for use in low noise amplifiers in certain frequency ranges. The noise contribution of the amplifier itself is stated in terms of its **noise figure** - the lower the noise figure, the quieter the amplifier. In order to achieve the manufacturer's specified noise figure, the active device must be operated at the specified DC voltage and current, and must also be presented with a specific source impedance which seldom coincides with that required for a conjugate match. It is possible to plot noise figure contours on a Smith Chart - these are circles that indicate the allowable zone of source impedance that will result in a certain noise figure. A low noise amplifier design usually involves trading off gain for noise figure.

Earlier in this section we discussed the Rollett Stability Factor, and the condition for an active device to be unconditionally stable. We avoided the use of a transistor that had a K of less than 1.0 because of the possibility of oscillation at certain combinations of source and load impedance. Although this is a very conservative precaution, there is no reason to avoid devices with a K of less than 1.0 if you are careful in your choice of source and load impedance. It is possible to plot **Stability Circles** on a Smith Chart - these are mathematically generated circles that outline the range of source and load impedances that lie within the stable region and the unstable region. The use of these Stability Circles allows the designer to choose impedances that will result in a stable design.

Modular Amplifier Components

For general purpose RF amplifier needs, an easy solution is commercially available that does not require any matching design at all! Several companies (HP and Mini-Circuits are the most common) supply inexpensive, small RF amplifiers with internal RF matching networks. One of the most common of the packages is approximately one quarter inch diameter, and about one eighth inch thick. Two ground leads are provided, plus an input and an output. Other packages include surface mount and flat pack. The input and output impedances are close enough to 50 ohms that external matching seldom needs to be done. The designs are inherently stable, and they can be easily cascaded to produce the desired overall gain.

A sample of just some of the devices that are available is as follows:

Mini-Circuits

Model	Frequency Range (MHz)	Gain (dB)	Maximum Output Power (dBm)
MAR-1	0-1000	13	+1.5
MAR-2	0-2000	8.5	+4.5
MAR-3	0-2000	8.0	+10.0
MAR-4	0-1000	7.0	+12.5
MAR-6	0-2000	9.0	0.0
MAR-7	0-2000	8.5	+4.0
MAV-1	0-1000	12.5	+1.5
MAV-2	0-1500	7.5	+4.5
MAV-3	0-1500	7.5	+10.0
MAV-4	0-1000	7.0	+11.5
RAM-1	0-1000	13	+1.5
RAM-2	0-2000	8.5	+4.5
RAM-3	0-2000	8.0	+10.0

RAM-6	0-2000	9.0	+2.0
RAM-7	0-2000	8.5	+5.5
RAM-8	0-1000	19.0	+12.5
VNA-25	500-2500	11.5	+15.5

HP *(Note that the HP semiconductor division was "spun off" to Agilent, and this division was then sold and morphed into Avago, which now sells similar devices with different part numbers)*

Model	Frequency Range (MHz)	Gain (dB)	Maximum Output Power (dBm)
MSA-0104	0-800	17.0	+1.5
MSA-0135	0-1200	18.0	+1.5
MSA-0185	0-1000	15.5	+1.5
MSA-0186	0-900	17.5	+1.5
MSA-0204	0-1800	10.0	+4.5
MSA-0235	0-2700	11.5	+4.5
MSA-0285	0-2600	10.0	+4.5
MSA-0286	0-2500	10.0	+4.5
MSA-0304	0-1600	10.0	+10.0
MSA-0335	0-2700	11.5	+10.0
MSA-0385	0-2500	10.0	+10.0
MSA-0386	0-2400	10.0	+10.0
MSA-0404	0-2500	7.0	+11.5
MSA-0435	0-3800	7.5	+12.5
MSA-0485	0-3600	7.0	+12.5
MSA-0486	0-3200	7.0	+12.5
MSA-0504	0-2300	16.0	+18.0
MSA-0505	0-2300	16.0	+18.0
MSA-0685	0-800	17.0	+2.0
MSA-0986	0-5500	6.0	+10.5
MSA-1023	0-2500	7.5	+25.0
MSA-1104	0-1300	10.0	+17.5
MSA-1105	0-1300	10.0	+17.5

NEC *(Available through California Eastern Laboratories)*

Model	Frequency Range (MHz)	Gain (dB)	Maximum Output Power (dBm)
UPC2710TB	50-1000	33	+10.8
UPC3224TB	50-3200	24	-3.5
UPC2708TB	50-2900	15	+9.2

All of these amplifiers are wideband devices. For narrow band operation, the units should be interfaced to appropriate filtering circuitry. These amplifiers require external coupling capacitors and biasing resistors. The designer should refer to the manufacturer's data sheets before using any of these devices. Although easy to use, care must be taken in grounding and biasing these components. Many of these amplifiers can be purchased for just a few dollars each in even in small quantities, making for a very cost effective design solution.

RF LARGE SIGNAL AMPLIFIERS

The previous section dealt with RF amplifiers operating in Class A - the active device was passing current for the entire period of the RF waveform. This gives excellent fidelity, but is not overly efficient. Here we define efficiency as the ratio of the amplifier's RF output power to the DC power input. For small signal RF amplifiers such as those in receivers, efficiency is not usually an issue because of the very low output powers involved. When the required amplifier output power exceeds about +20 dBm (100 milliwatts), efficiency needs to be considered because of power supply requirements and component heating.

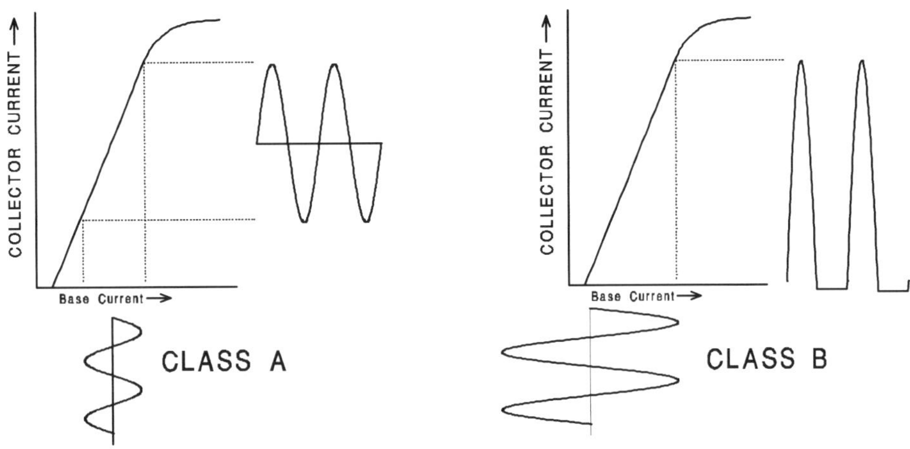

The active device in a Class A amplifier conducts over all 360 degrees of a cycle of the RF waveform, and passes the same average DC current whether a signal is present or not. A Class B amplifier is biased so that it only conducts for 180 degrees, and it only draws current when a signal is present. Obviously, large amounts of distortion are created if only half of the RF cycle is being amplified! Class B is usually used in "**push-pull**" amplifiers - one active device amplifies the positive part of the waveform, and one amplifies the negative part:

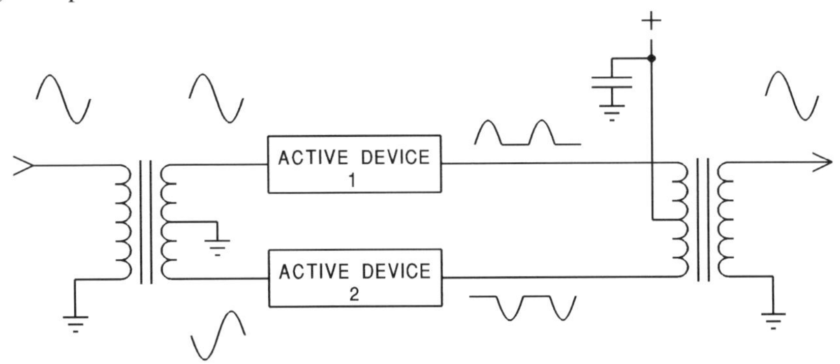

This same circuit has been used for many years in tube-type Hi Fi power amplifiers.

A push-pull Class B amplifier is much more efficient than a Class A amplifier, but it suffers a small amount of distortion as the signal goes through zero - one device is just ceasing conduction while the other one is just about to start conduction. This is called "**zero-crossing distortion**", and it can be minimized by adjusting the bias levels so that a small amount of current flows through both active devices in the absence of an input signal. For very small input signal levels, our push-pull amplifier now operates as a Class A design, but larger amplitude signals are amplified in Class B. This is called **Class AB**. Class AB amplifiers are used as linear power amplifiers from audio through RF frequencies.

A linear amplifier such as Class A or AB faithfully reproduces input amplitude variations at the output. If amplitude modulation in any form is on the input signal, a linear power amplifier is needed in order to ensure that the modulation is still present on the output signal.

Higher efficiency RF power amplifiers can be designed by biasing the active device so that it conducts for less than 180 degrees of the input waveform's cycle. These **Class C** amplifiers produce lots of harmonic distortion, and they can't be used as a linear amplifier. In order to suppress unwanted harmonic outputs, and also to create an output waveform resembling a sine wave, the output tuning circuit must have adequate Q. A minimum value of 10 for the Q of the output tuning network means that a significant amount of energy is stored in the reactive components - this allows the tuned output network to supply energy to the load during that period of the cycle when the active device is cut off - in other words, the output tuned circuit (typically called the "tank circuit") acts like a flywheel.

Carefully designed Class C amplifiers can achieve efficiencies approaching 70%. If amplitude modulation (AM or SSB) is used, the modulation must be applied after the Class C amplifier because of its non-linear transfer characteristic. In the old days of AM transmitters on the HF bands, it was common to apply the audio modulation to the positive supply of the final Class C RF amplifier.

An FM signal does not contain amplitude modulation components, so it can be amplified effectively by a Class C amplifier.

The design approach to a transistor RF power amplifier is similar to that used for small signal amplifiers for the input matching network. The output circuit is different however. The matching network should be designed so that the resistive part of the load impedance presented to the active device's output is equal to:

$$R_L = \frac{(V_{CC} - V_{SAT})^2}{2P} \text{ ohms}$$

where R_L = resistive part of the load impedance
V_{CC} = power supply voltage
V_{SAT} = the saturation voltage of the active device (typically 1 or 2 volts for a BJT)
P = the desired output power in Watts

Once the resistive part of the load has been matched as above, reactive components (either capacitive or inductive) can be used to null out any residual reactance.

Heat dissipation is a big consideration in the design of an RF power amplifier. If a 250 Watt Class C amplifier has an efficiency of 60%, 166 Watts of power will be dissipated as heat in the active device. Tube amplifiers can dissipate this heat easily with forced air cooling, but transistor and FET power amplifiers need something more than a cool breeze over the package! Semiconductor devices are physically small, and their surface area is insufficient for dissipating much more than a watt via convection cooling to the surrounding air. Power FETs and transistors are made with a stud or mounting flange that is designed to conduct heat into a heat sink. Manufacturer's data can be used to determine the maximum allowable temperature on the stud or flange - it is a function of the power dissipated in the semiconductor's junction and the thermal resistance between the junction and the stud or flange. The heat sink may be a large finned aluminum extrusion in an airflow, or for a high power design it might be a copper block with cooled water circulating through internal passages.

It is often convenient to implement high power amplifiers by using combinations of lower power amplifiers. If this is done, power splitters and power combiners should be used to ensure equal splitting of the power generation.

A **Class D** amplifier is analogous to a switching power supply. The active components are either turned off or are "hard on". Very little power is dissipated in the active components, and efficiencies are high. At high frequencies, it gets harder to cause a transistor or FET to turn "on" or "off" quickly enough, and power is dissipated during the transitions - this creates heat. Class D amplifiers are commonly used in high power audio amplifiers, but only a few specialized commercial communication companies build expensive Class D amplifiers for RF broadcast frequencies.

Modular RF power amplifiers are available from several commercial sources. Motorola, Toshiba, and Mitsubishi are prolific suppliers, and offer a selection of component amplifiers covering the power range up to about 20 Watts and beyond 1 GHz.

TRANSMITTERS

A radio transmitter is simply an amalgamation of many of the components that we have already discussed into a single package. We will illustrate this by looking at a typical block diagram for an FM VHF transmitter:

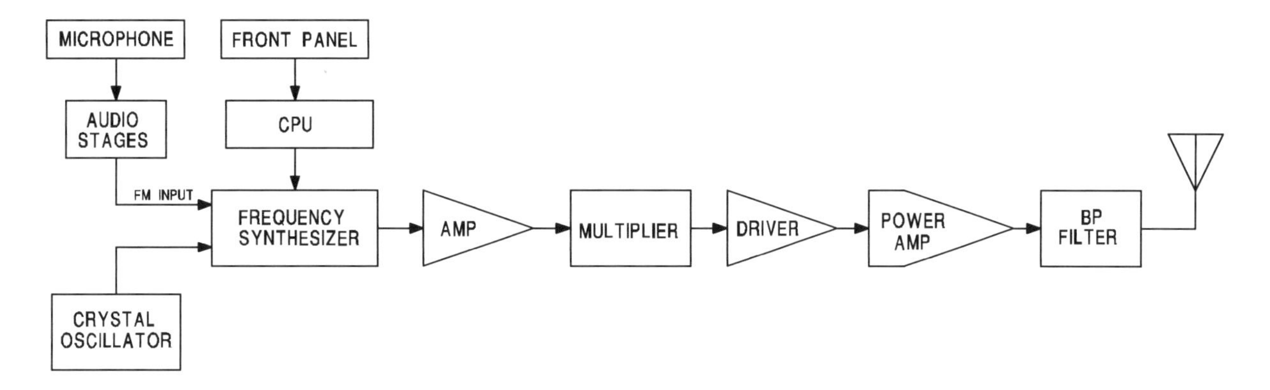

We will briefly look at each of the blocks.

The microphone creates a small audio voltage which is amplified and filtered in the audio stages. The resulting voltage is fed into the frequency synthesizer where it is applied to the control input of the VCO in the synthesizer's PLL to create FM.

The front panel contains all of the push buttons, knobs, and displays that interface with the operator. Since this is a modern transmitter, a small microprocessor block (called the "CPU") interfaces with the front panel to provide a variety of memory features and operator conveniences. The CPU sets the frequency synthesizer's output frequency by programming the divide ratios in the PLL's frequency dividers. The PLL uses a reference frequency that is derived from the crystal oscillator.

The output of the synthesizer consists of a low amplitude NBFM signal. It is amplified, then multiplied up to the desired output frequency. The modulation index is increased in the multiplier. A stage called the "driver" boosts the power level to the point where it is sufficient to drive the Class C output power amplifier. Harmonics of the amplifier's output are suppressed by the output bandpass filter before the signal is fed to the antenna.

Single Sideband

A modern HF SSB transmitter is more complicated. We will examine the blocks in a hypothetical design. The internal frequencies used do not match those of any specific piece of commercial equipment, but were simply chosen to show one approach to the design:

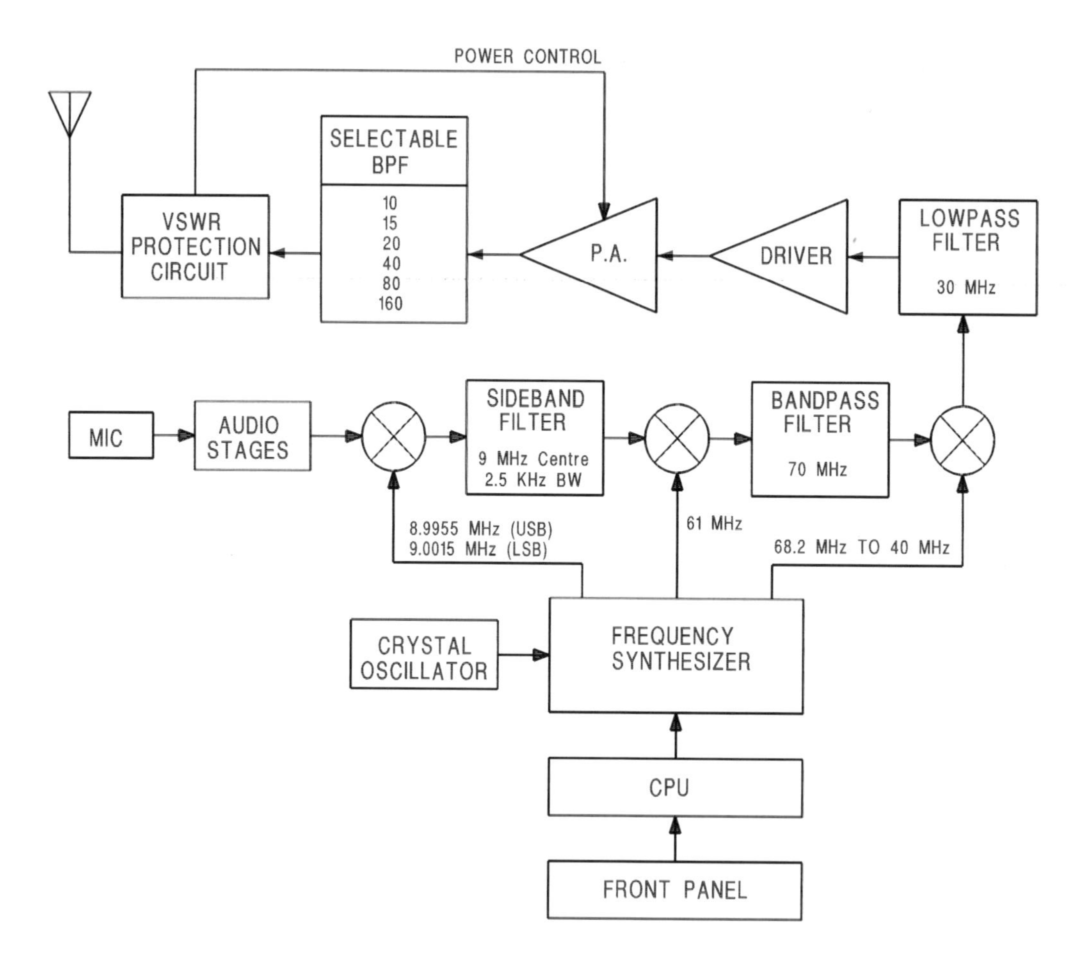

A CPU is again used to provide the interface with the transmitter's front panel. The synthesizer block actually contains circuitry to generate several different output frequencies that are all derived via PLL's and dividers from the crystal oscillator.

The audio is amplified and fed to a balanced mixer which creates a double sideband signal at approximately 9 MHz. A sharp bandpass filter (probably built using multiple crystals) is used to select one of the sidebands. The carrier frequency into the balanced mixer is placed either above or below the centre frequency of the sideband filter so that the filter's pass band will be centred on either the upper or lower sideband. The output of the sideband filter is mixed with a fixed 61 MHz signal and followed with a bandpass filter so that the SSB signal is now centred on 70 MHz.

As the operator tunes the desired output frequency (in the range of 1.8 MHz to 30.0 MHz) using the front panel controls, the CPU commands one of the outputs of the frequency synthesizer to tune within the range of 68.2 MHz to 80 MHz. This frequency is applied to one input of a mixer, together with the previously-generated 70 MHz SSB signal. A low pass filter then selects the desired difference frequency range of 1.8 MHz to 80 MHz.

A driver and linear power amplifier (probably Class AB) boost the signal to the desired output power level. A bandpass filter centred on each band of interest is used to suppress unwanted output signals: the correct filter is switched in by the CPU. Since this example is of am amateur transmitter, there is one filter for each of the bands (10 through 160 metres).

A VSWR protection circuit monitors the reflected power from the load. If it exceeds a specified level, the gain of the power amplifier is reduced to avoid damaging any components. The VSWR protection circuit contains a broadband directional coupler that is used to sense the reflected power.

RECEIVERS

A receiver can be considered as being no more than a tuneable filter, an amplifier, and a demodulator. Indeed, the earliest radios dispensed even with the amplifier!

The object is to amplify the selected signal and demodulate it while ignoring all of the other signals that may be producing an input at the antenna. The desired signal may be very weak or quite strong, and it may be on a frequency that is quite close to other, strong interfering signals. The desired signal's amplitude may be as small as 0.5 μV, or it might be as strong as 1 volt - this implies a dynamic range of 126 dB!

The bandwidth of the receiver should be just sufficient for the particular operational mode used: 10 KHz for AM, 5 KHz for NBFM, about 2.5 KHz for SSB, and perhaps 500 Hz for CW. If a receiver were designed to receive CW on 2 metres, this would imply that the effective Q of the receiver's filtering capability would be 144,000/0.5 = 288,000! It is virtually impossible to design a filter with this sort of Q for even a single frequency, so it is clear that some other approach should be taken in a receiver.

Sticking with analogue approaches, the answer to the bandwidth dilemma is to use a superheterodyne receiver design. All input RF frequencies will be down-converted to a lower intermediate frequency (the "**I.F**".) where the selective filters will be easier to implement with a more reasonable value of Q. As an example, if the I.F. is at 455 KHz, a 500 Hz bandwidth would imply a required filter Q of 900 - this could be easily achieved with a ceramic filter or multiple LC filters. A tuneable local oscillator (the "**L.O.**") is used to mix the desired frequency down to the intermediate frequency in the first mixer. The local oscillator's frequency can be either higher or lower than the input frequency by an offset equal to the I.F. It should be obvious that the superheterodyne receiver will therefore respond to two different frequencies - one above the L.O. frequency, and one below it. This undesired response is called the "**image response**".

In order to eliminate the image response, some form of selectivity is needed between the antenna and the first mixer. This is usually obtained in the form of a bandpass filter. If the filter is narrow, its centre frequency will have to track any changes in the L.O. frequency.

A basic block diagram of a simple superheterodyne receiver is shown below:

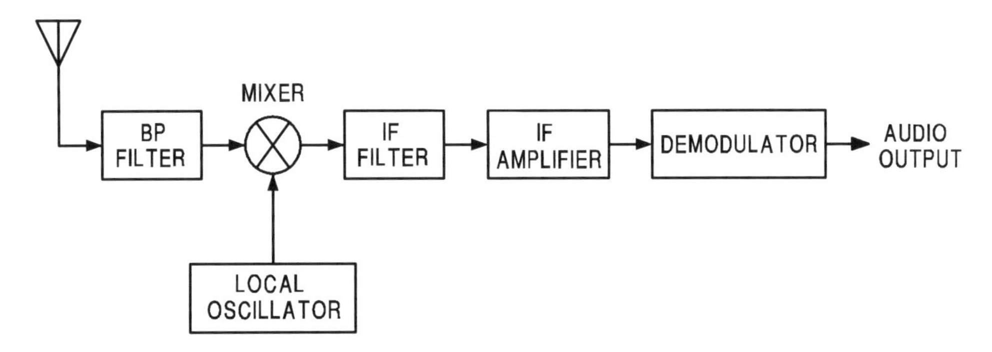

In order to achieve adequate selectivity, a low IF frequency should be used. Unfortunately this results in the image response being fairly close to the desired frequency. If a 455 KHz IF is used, the image response is only 910 KHz away. If you are trying to listen to a 1 microvolt signal at the same time that there is an undesired 1 volt signal on the image response frequency, the requirements for the band pass filter in front of the mixer are going to be extremely severe! One solution to this problem is to use **double conversion** - the first IF is quite high in order to satisfy the image response concerns, while the second IF is fairly low in order to achieve adequate selectivity. Modern receivers sometimes use even triple and quadruple conversion.

The local oscillator in receivers used to be a continuously-tuneable analogue unit - the design objective was to design a stable oscillator that had a fairly linear frequency change per unit change of the tuning dial. In the mid 70's the digital revolution overtook receiver design, and almost all modern receivers use frequency synthesizers to generate the L.O. frequency or frequencies. This provides the advantage of stability, settability, and memory features. In order to give the operator the "feel" of an analogue tuning system, the receiver's tuning knob is equipped with a flywheel and friction brake. The knob operates an optical shaft encoder that increments or decrements a digital register defining the receive frequency. As long as the frequency steps are less than about 20 Hz, the receiver "feels" just like one of the good old analogue units.

The following block diagram illustrates a hypothetical 1 to 30 MHz receiver designed using modern approaches:

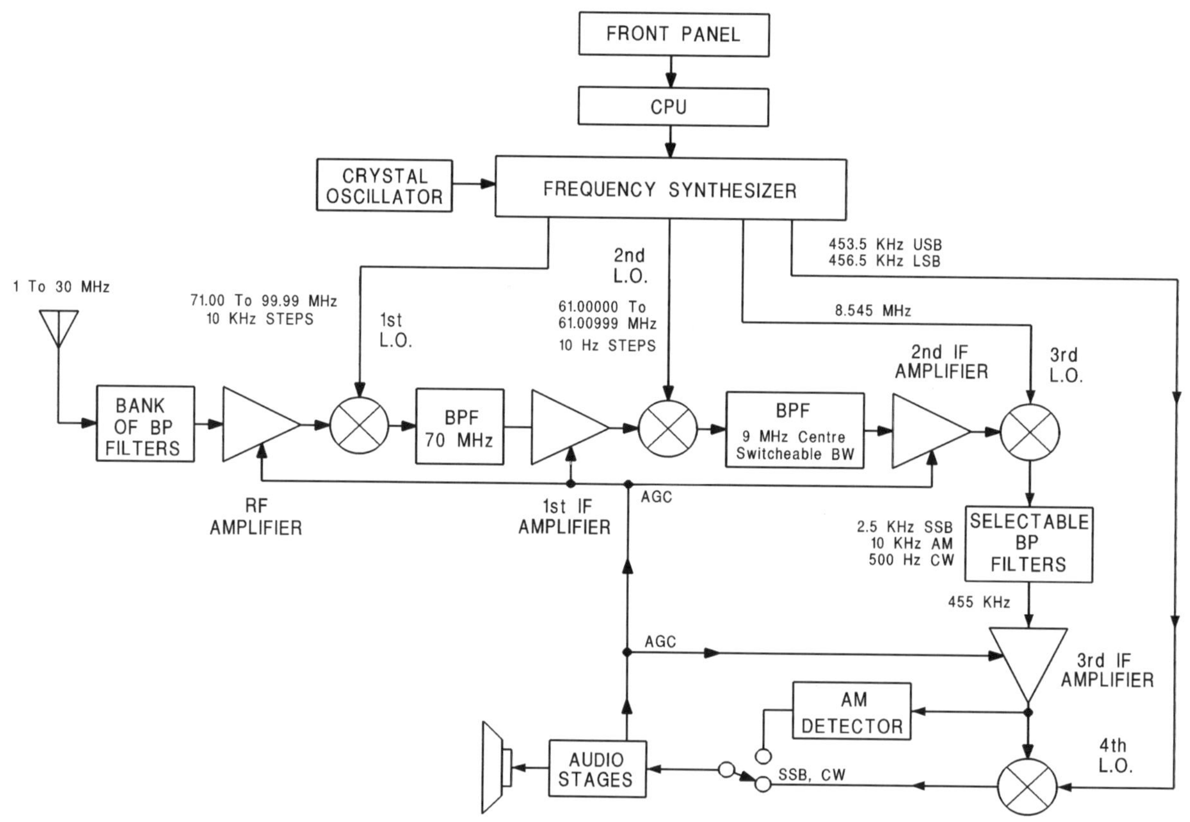

This particular block diagram and the frequencies chosen do not necessarily match any commercially available equipment, but has been structured to illustrate a number of modern principles. The receiver uses quadruple conversion for SSB and CW signals, and triple conversion for AM.

The first IF is at 70 MHz, so the image response is 140 MHz higher than the desired frequency. A simple 30 MHz low pass filter before the 1st mixer would suffice to eliminate almost all image response concerns, but this receiver uses a bank of switched band pass filters in the front end - this reduces the possibility that a very strong signal at a frequency far removed from the desired frequency will cause an overload condition of the RF amplifier.

Very little selectivity is provided at the first IF, and indeed the filter must be wide enough to accommodate not only the widest transmission to be received, but also the fact that the 1st IF actually varies over about a +/- 5 KHz range.

The 2nd IF at 9 MHz has a series of selectable filters. These are most likely implemented with multiple crystals. The 3rd IF at 455 KHz is where the serious filtering occurs.

AM signals are demodulated by an envelope detector that operates at 455 KHz. SSB and CW signals are demodulated by mixing in a product detector with the 4th L.O. from the synthesizer, which emulates the BFO in a more traditional receiver.

Because of the wide dynamic range of allowable input signals, the gain of the RF amplifier and all of the IF amplifiers is varied in accordance with the perceived signal strength as sensed in the final audio stages. This is called Automatic Gain Control (AGC). It is normal for the AGC time constants to be chosen so that the gain is rapidly reduced for an increasing signal strength, but slowly increases for reducing signal strength.

Noise

Receivers are often used to listen to extremely weak signals. Even in the absence of man-made interference, the signal will be contaminated with noise, and the operator will often have to strain to make out the message content. The noise that is heard coming from the audio stages of a receiver is a combination of the atmospheric noise and the receiver's own internal noise. Obviously it would make sense to minimize the internal noise contributions as much as possible, because they define the ultimate sensitivity of the receiving system.

Any electronic component produces noise that is caused by random electron flow within it. The magnitude of the noise is proportional to the absolute temperature of the component. This is why radio telescopes often cool the front end components to extremely low temperatures in an attempt to discern extremely weak signals. The thermal noise voltage developed across any conducting component is as follows:

$$V = \sqrt{4kTBR}$$

where k = Boltzmann's Constant (1.38×10^{-23})
T = Absolute temperature in degrees K
R = resistance of the device in ohms
B = bandwidth in Hz

Active devices like transistors or FETs exhibit noise temperatures that are different from the ambient temperature. The noise contribution of amplifiers is sometimes specified in terms of its **noise temperature**. Theoretically this is determined by placing a terminator on the input of the amplifier and cooling it to absolute zero (rather hard to do!). The temperature of the terminator is then increased until the noise output power of the amplifier has increased by 3 dB - the terminator's temperature is then called the amplifier's equivalent noise temperature.

Low noise RF amplifiers are normally specified in terms of their **noise figure**. This is simply the dB ratio of the SNR power at the input of the amplifier to that at the output. This gives an indication of how much noise is added to the signal by the amplifier. In other words:

$$NF = 10 \log\left(\frac{\text{input SNR}}{\text{output SNR}}\right)$$

Software Defined Receivers

A new class of receivers has achieved popularity since about 2000: Software Defined Receivers (SDR). These receivers use the latest advances in very high speed, high resolution (greater than 16 bits) Analogue to Digital converters, together with high throughput Digital Signal Processors. Using this type of architecture, the receiver's frequency, bandwidth, and demodulation type can all be defined by software running on the DSP. SDR technology is often used in combination with a more traditional analogue superheterodyne architecture - SDR processing starts at the first I.F. amplifier.

The conversion from analogue to digital is done with two very high speed A/D converters, configured such that digital samples of the input waveform are taken both "in phase", and at a 90 degree phase shift. The two streams of digital values from the A/D converters are labeled as "I" and "Q" (these are abbreviations for "In-Phase", and Quadrature). The I and Q data streams are processed, and any type of demodulation and filtering can be performed in software. The rate of sampling and A/D conversion must exceed twice the highest input frequency of interest (due to the Nyquist Sampling Theorem). It is even possible to have multiple "virtual receivers" defined in software that can receive multiple frequencies at once!

Theoretically, the ideal SDR would take the antenna signal and feed it directly into the samplers and their associated A/D converters, eliminating all forms of analogue processing. A Low pass filter would be required before the samplers to ensure that only input signals below the Nyquist Sampling Limit (which is one half of the sampling frequency) are passed. In practice, the limited dynamic range and sensitivity of current A/D converters will require some type of analogue RF amplification before the conversion.

New modulation methods (either analogue or digital) can easily be added to an existing SDR with just a software update! Much of the SDR hardware can also be used to digitally generate signals as part of a software-defined transmitter.

The military were one of the earliest adopters of SDR equipment, but there are now several manufacturers of software-defined radios targeting amateur radio users. Groups of volunteer amateur designers have also designed some impressive SDR equipment, and is sharing their expertise as "shareware", The interested reader is encouraged to look at organizations such as OpenHPSDR and GNUradio, and also vendors such as Apache Labs and FUNcubeDongle to see some of the exciting developments in this area.

The detail design of SDR equipment is beyond the scope of this publication!

Intermod

In today's crowded RF environment, problems due to intermodulation distortion (commonly referred to as "intermod") seem to be afflicting us more and more. You have probably already heard local paging signals coming out of your VHF radio as you drive by areas that have lots of commercial antennas visible - this is due to intermod.

Thirty years ago, intermod problems were much less severe. This is due to the fact that there was much less RF "floating around" then, but also because older VHF or UHF ham rigs had much "tighter" front ends - they were meant for receiving the ham bands only, and did not have broadband coverage. Most of today's modern wide coverage receivers are quite susceptible to intermod problems in metropolitan areas.

As described above, most modern receivers are superheterodyne configurations. The incoming RF signal is applied to one input of a mixer, and a local oscillator is applied to the mixer's other input. The local oscillator frequency is chosen so that it differs from the desired incoming signal by an amount equal to the intermediate frequency (I.F.), which is commonly (for popular hand-held VHF FM radios) in the range of 45 MHz. The local oscillator can be higher or lower than the incoming signal - all that matters is that it differ by an amount equal to the I.F. frequency. All of the receiver's selectivity is achieved by filters at the I.F. frequency. The following block diagram will remind us of this architecture:

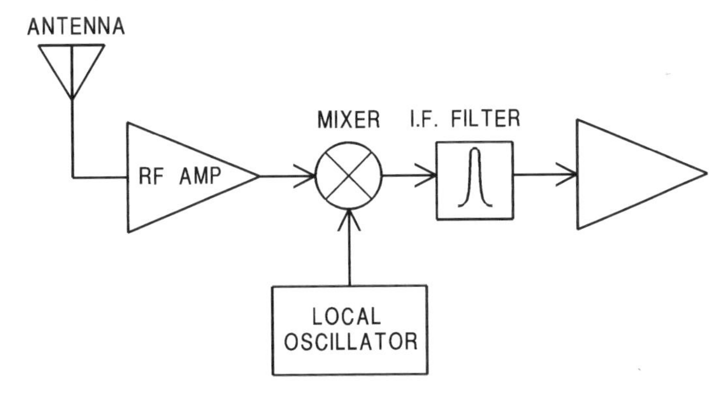

Assume for the moment that receiver is designed to work on the amateur 2 metre band (144-148 MHz), and that the I.F. is 45 MHz. If the local oscillator is set to a frequency that is 45 MHz below the desired signal, the rcceiver will have an image response 90 MHz below the desired signal (somewhere around 54 MHz). Because the image frequency is so far away from the band of interest, some simple filtering ahead of the mixer will suffice to attenuate the image response to very small levels. The front end filtering required to give decent image response is quite rudimentary, and there is no requirement that the filter's centre frequency be varied as you tune from end to end on the 2 metre band.

If all of the signals that are on the air are contained within the 2 metre band, the above-described superheterodyne receiver will work just fine. Unfortunately, there are a myriad of other signals that will be picked up - paging, police, taxi, broadcast, etc. If these signals are within 10 or 20 MHz of the 2 metre band, they will show up at the mixer's input with very little attenuation. A mixer is a non-linear device, and the output of it will contain sum and difference frequencies of all the signals with respect to each other, plus sums and differences of their harmonics.

As an example of a problem frequency combination, consider an FM broadcast station on 107 MHz, and a pager on 152 MHz. If both signals are present at the input to the mixer, a 45 MHz output will be produced that will pass right through the receiver's I.F. filter. The received audio under this circumstance will contain modulation from both signals. A modern wide coverage ham receiver will have little selectivity before the mixer, and the FM broadcast signal will receive only moderate attenuation, and the pager may not be attenuated at all! If the receiver is located in an area where these signals are strong enough, a spurious response due to intermod will result. Certain areas of cities are notorious for intermod problems - these "intermod alleys" are bombarded by RF from a multitude of sources, and the frequency combinations can cause problems if a sum or difference lands on the I.F. frequency. Different receivers with different I.F.'s or different front end filtering will behave differently.

The only sure fire way to solve intermod problems is to place sufficient selectivity ahead of the mixer such that problem combinations of frequencies cannot simultaneously pass through. Older narrow band 2 metre receivers contained filters that exhibited strong attenuation outside of the 144 to 148 MHz range. Top of-the-line commercial receivers used helical filters that give extremely good selectivity. Modern broad coverage amateur receivers can be "cured" of intermod problems by placing a band pass filter in line with the antenna. The use of an external filter will, however drastically decrease sensitivity to out of band signals (such as police or fire) that you may actually wish to listen to.

The **real** answer is to design the receiver with a narrow filter in the RF section (before the mixer) that tracks the receive frequency. This is fairly easy to do by using a varactor diode to tune the filter's centre frequency in synchronization with the local oscillator. At least one manufacturer of upscale amateur 2 metre equipment has such a feature.

Rather than having the front end filter be continuously tuned, it is possible to switch in different filters to cover bands of frequencies. The filters do not have to be razor sharp, but the filter bandwidth must be less than the I.F. frequency.

VHF and UHF Repeaters can also suffer from intermod. They are usually located in very close proximity to many other repeaters, and very high unwanted signal levels are often present at the input to the receiver's mixer. Even the use of cavity filters cannot totally prevent sporadic intermod from occasionally tripping some machines. In order to minimize this problem, repeaters often are configured to require the use of a sub-audible tone, such as 100 Hz. This means that the repeater receiver's squelch circuit is triggered by the presence of the 100 Hz tone, not by the perceived carrier level. The probability that random intermod will result in an extraneous 100 Hz audio modulation is slight, and the machine mostly ignores all of its noisy neighbours!

APPENDIX A - REVIEW OF THE BASICS

It is assumed that much of the following summary information was already familiar to the reader before reading this text -

Ohms Law: $\frac{E}{I\,R}$ (cover the unknown, look at the resulting relationship)

$P = E \times I$ Watts

$P = I^2 R$ Watts

$P = \frac{E^2}{R}$ Watts

Resistors in series: $R_{total} = R_1 + R_2 + R_3 + ...$

Resistors in parallel: $R_{total} = \frac{1}{\frac{1}{R_1} + \frac{1}{R_2} + \frac{1}{R_3} + ...}$ ohms

Or, for two resistors: $R_{total} = \frac{R_1 \times R_2}{R_1 + R_2}$ ohms

Conductance: $G = \frac{1}{R}$ mhos (also called Siemens)

Capacitive reactance: $X_C = \frac{1}{2\pi f\, C}$ ohms

Inductive reactance: $X_L = 2\pi f\, L$ ohms

Resonance: $f_{res} = \frac{1}{2\pi\sqrt{LC}}$ Hz. (Note that at resonance, $X_C = X_L$)

or: $\omega_{res} = \frac{1}{\sqrt{LC}}$ rad/sec.

Susceptance: $B = \frac{1}{X}$ mhos (or Siemens)

Wavelength: $\lambda = \frac{c \times v}{f}$ metres,

where: c = speed of light in free space (~300,000,000 m/sec)
v = velocity factor (1.0 for free space)
f = frequency in Hz.

Half Wave Dipole: $\text{Length} = \frac{468}{f_{MHz}}$ feet

Transformers: $V_{out} = \frac{V_{in}}{N}$, $Z_{out} = \frac{Z_{in}}{N^2}$ where N is the turns ratio

APPENDIX B - REACTANCE CHARTS

If both frequency and reactance are plotted on logarithmic scales, the resulting log-log graph has very interesting properties. Using a chart of this type, plots of reactance versus frequency are straight lines for both capacitors and inductors. It is easy to determine resonant frequencies for any C and L combination just by looking to see where the plots intersect. The following charts are very useful in analyzing AC circuits:

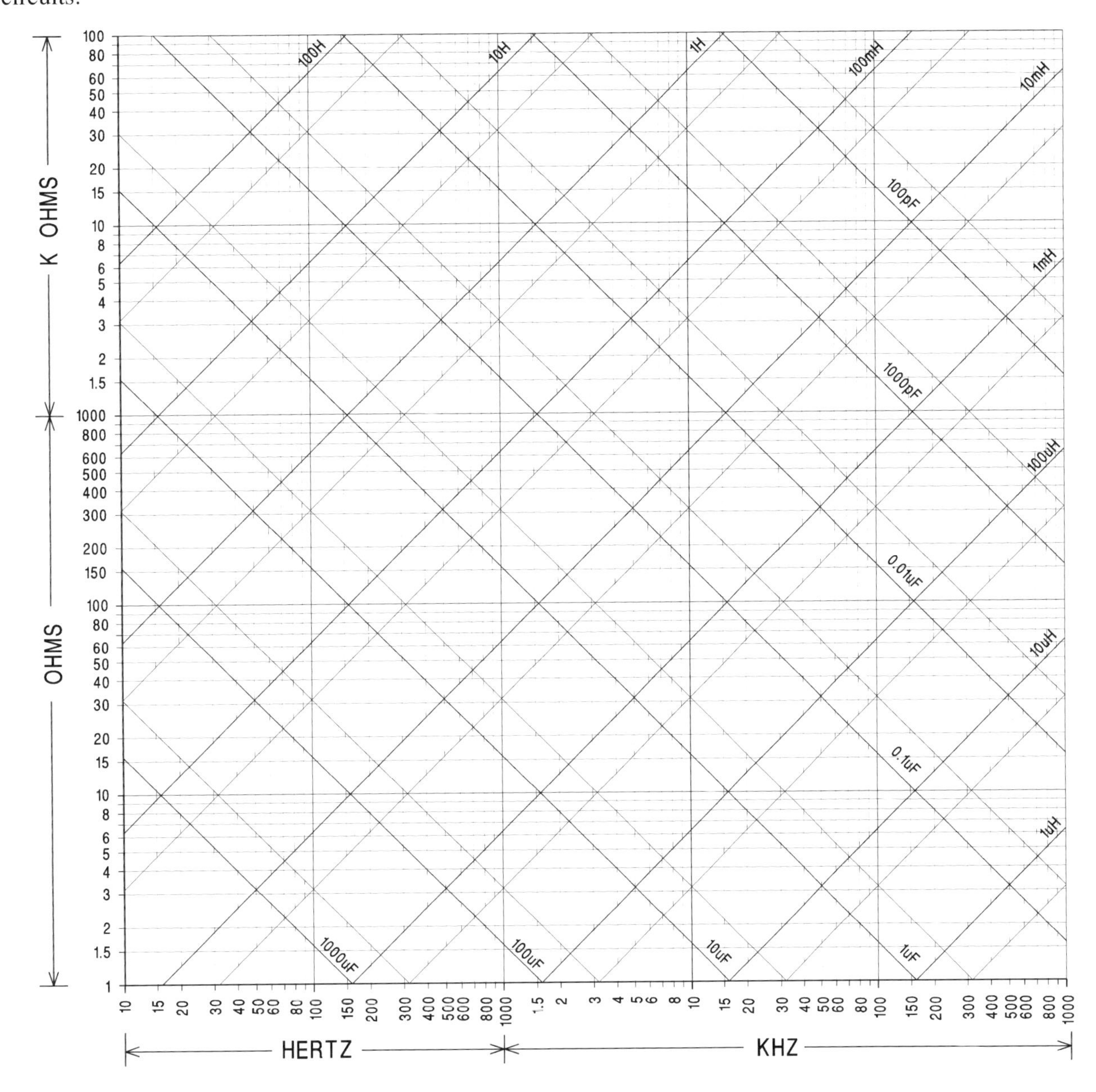

REACTANCE CHART - Audio Frequencies

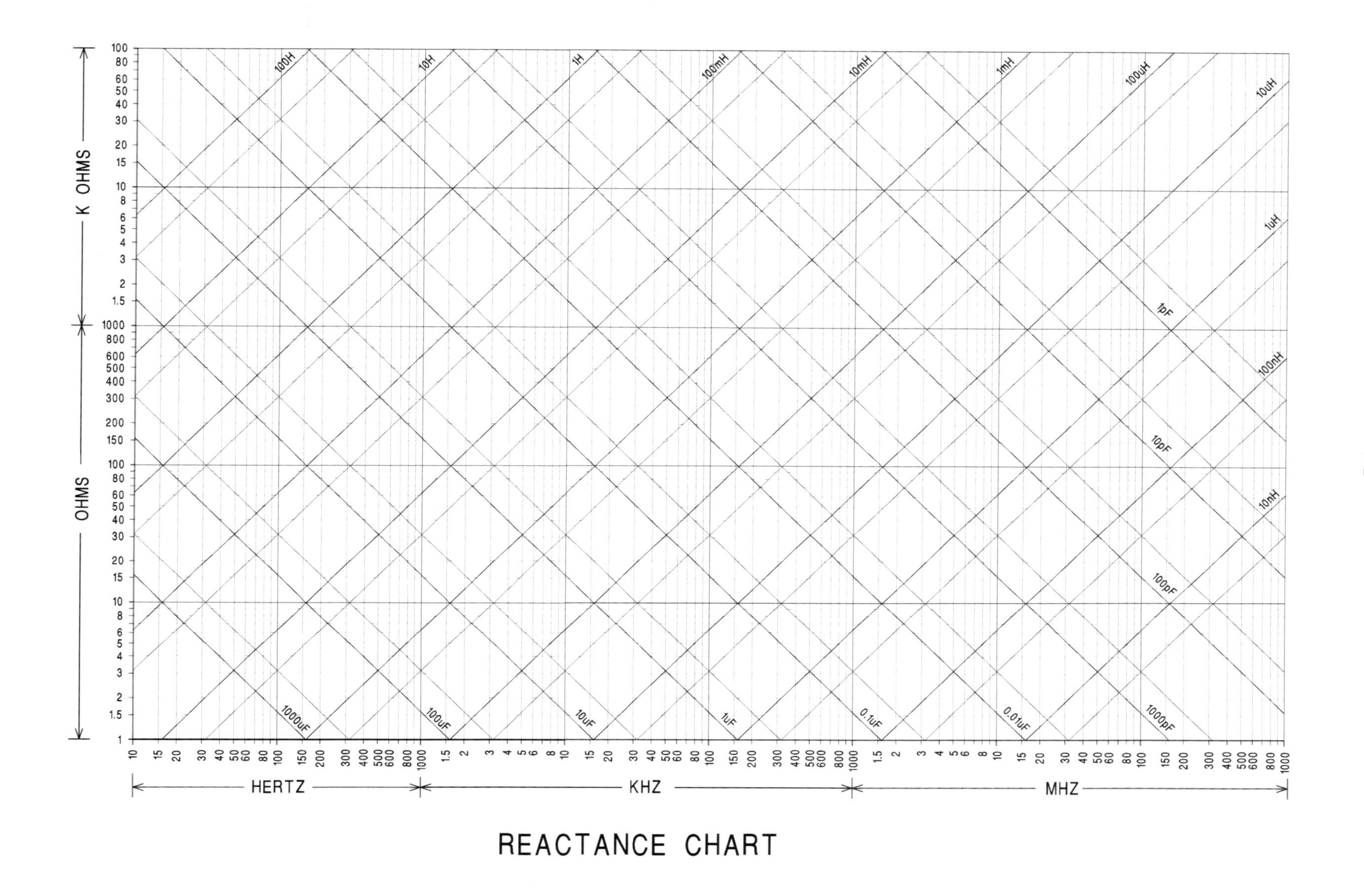

REACTANCE CHART

APPENDIX C - SMITH CHART

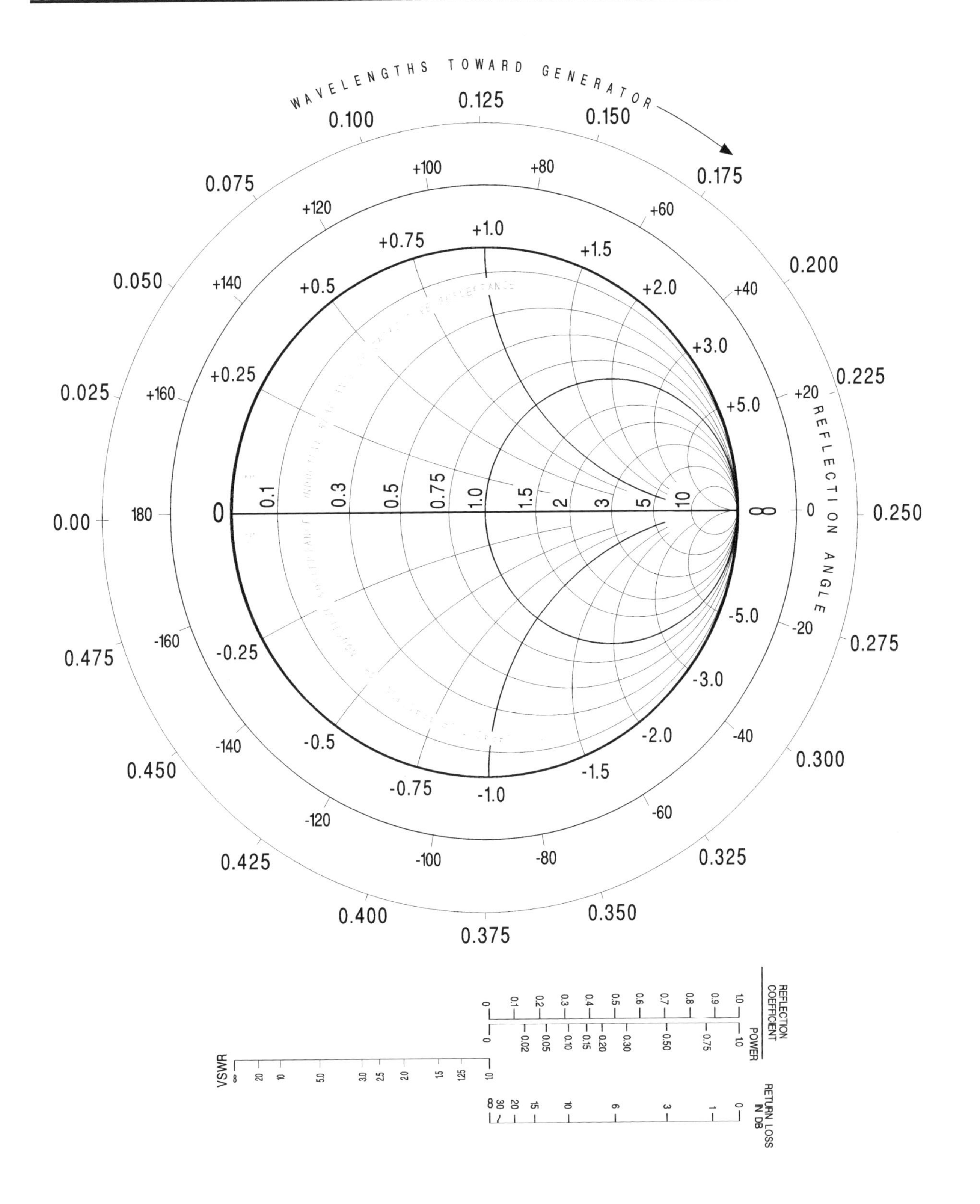

APPENDIX D - USEFUL INFORMATION

Much of the following information was discussed in the text, while other information is presented in the spirit of completeness.

Impedance: $Z = R + X$, or $|Z| = \sqrt{R^2 + X^2}$ ohms

Admittance: $Y = \frac{1}{Z}$ mhos (or Siemens)

Admittance: $Y = G + B$, or $|Y| = \sqrt{G^2 + B^2}$ mhos (or Siemens)

Converting from Y to Z: $R = \frac{G}{G^2 + B^2}$, $X = \frac{-B}{G^2 + B^2}$.

Converting from Z to Y: $G = \frac{R}{R^2 + X^2}$, $B = \frac{-X}{R^2 + X^2}$

Quality Factor: For inductors, $Q = \frac{X_L}{R_L}$ For capacitors, $Q = \frac{R_C}{C_C}$ where R_L is the inductor's series resistance, and R_C is the capacitor's parallel resistance. Note that Q is frequency dependent.

Q of a tuned circuit: $Q = \frac{f_{centre}}{BW_{3dB}}$

Forward-biased junction: V_{BE} decreases approximately 2.1 mV/degree C. At a constant temperature, the current increases approximately ten fold for every 60 mV increase in V_{BE}.

Reverse-biased junction: Leakage current doubles for every 10 degree C increase.

Power: 1 BTU $\cong$ 1005 Watt-Seconds

Angular Measure: 1 revolution = 360° or 2π radians

1 radian $\cong$ 57.3°

Solid Angles: $\frac{\text{Area}}{r^2}$ steradians (abbreviated "sr")

Solid Angle of a Sphere: 4π sr.

AC Voltages and Currents: $V_{RMS} = \frac{V_{peak-peak}}{2\sqrt{2}} = \frac{V_{peak-peak}}{2.828}$

RMS voltage: That voltage which has the same "heating power" as an equivalent DC voltage.

AC Power: $P_{AC} = V \times I \times \text{Power Factor}$ watts

PF = cosine of the angle between current and voltage

Wire Inductance:

$$L = 0.002\,A\left[2.3\log\left(\frac{4A}{d} - 0.75\right)\right]\ \mu H$$

where A is the length of the wire in cm.
d is the wire diameter in cm.

1 inch of #20 wire has an inductance of about 7.5 nH.

Transmission line impedance: $Z = \sqrt{\frac{L}{C}}$ ohms. (where L and C are measured per unit length)

Free Space:

$\mu_0 = 4\pi$ E-7 henries/metre (permeability of free space)

$\varepsilon_0 = \frac{1}{36\pi}$ E-9 farads/metre (permitivity of free space)

Impedance of Free Space: $Z_0 = \sqrt{\frac{\mu_0}{\varepsilon_0}} = 377$ ohms

Reflections:

$\Gamma = \frac{E_r}{E_f}$ (reflection coefficient)

$\Gamma = \frac{R - Z_o}{R + Z_o}$

$|\Gamma| = \frac{VSWR - 1}{VSWR + 1}$

$VSWR = \frac{1 + \Gamma}{1 - \Gamma}$

Decibels:

dB = 20 log(V_2/V_1) (voltage ratios)
dB = 10 log(P_2/P_1) (power ratios)

dB_m = power relative to 1 mW into the same impedance
$dB_{\mu V}$ = voltage relative to 1 μV into the same impedance

0 dB_m = 0.2236 V_{rms} in a 50 ohm system

+3dB = power ratio of 2, or voltage ratio of $\sqrt{2}$ = 1.414
-3dB = power ratio of 0.5, or voltage ratio of $\frac{1}{\sqrt{2}}$ = 0.707

Logarithms:

e = 2.71828... (the base of natural logarithms)

if $x = 10^y$, then log(x) = y
if $x = e^y$, then ln(x) = y

APPENDIX E - BASIC TRIGONOMETRY

Trigonometry is a mathematical subject that deals with the relationships that exist in triangles. In particular, it deals with relationships between the lengths of the sides of triangles, and the angles at the three corners (called "vertices"). For this text, we will deal only with the most basic trigonometric functions.

A triangle has three sides, and three corners. Each corner subtends an angle, and the sum of the three angles is always 180 degrees in a triangle. It is common practice to arbitrarily label the length of the three sides as A, B, and C. The angles in the corners opposite these sides are referred to as a, b, and c. We can therefore say that a + b + c = 180 degrees.

We are mostly concerned with right angle triangles. These are triangles in which one of the angles is 90 degrees. We therefore know that the remaining two angles will add up to 90 degrees. Right angle triangles have relationships between the side lengths and included angles that have been tabulated and given special names such as **sine**, **cosine**, and **tangent**. These are abbreviated as sin, cos, and tan, and are defined as follows:

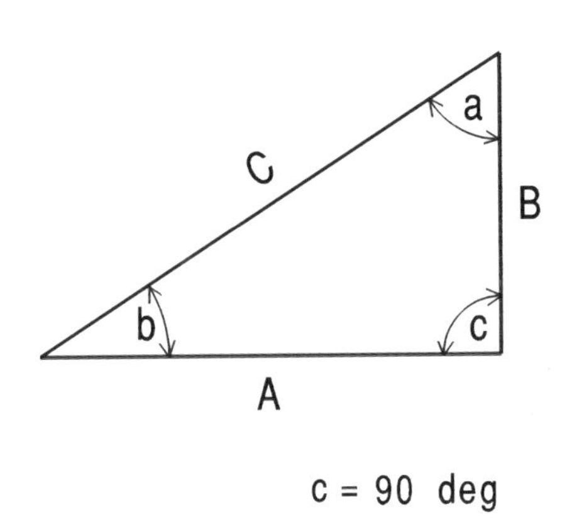

$$\sin(b) = \frac{B}{C} \qquad \sin(a) = \frac{A}{C}$$

$$\cos(b) = \frac{A}{C} \qquad \cos(a) = \frac{B}{C}$$

$$\tan(b) = \frac{B}{A} \qquad \tan(a) = \frac{A}{B}$$

The value of sin, cos, or tan can be determined for a given angle by looking up in published tables or by pressing the appropriate button on a calculator.

If angle b is equal to angle a, then they must both be equal to 45 degrees, and side A is the same length as side B. In this special case, sin(a) = sin (b) = cos(a) = cos(b) = cos(45). The values of sin, cos, and tan for some common angles are easy to remember:

sin (0) = 0	cos (0) = 1	tan (0) = 0
sin (30) = 0.5	cos (30) = 0.866	tan (30) = 0.577
sin (45) = 0.70707 ….	cos (45) = 0.70707….	tan (45) = 1
sin (60) = 0.866	cos (60) = 0.5	tan (60) = 1.732
sin (90)=1	cos (90) = 0	tan (90) = ∞

The famous theorem of Pythagoras relates to the length relationships in a right angle triangle. Looking at our illustration, the relationship is:

$$C = \sqrt{A^2 + B^2}$$

In looking at the illustration and some of the tabulated values, you will probably notice that for a given triangle, the sine of an angle is equal to the cosine of the other angle (that is not 90 degrees). In other words, using the illustration, sin(b) = cos(a).

There are some other interesting relationships that are not immediately obvious. We won't necessarily be using any of the following expressions, but they are included for the sake of completeness:

$$A^2 = B^2 + C^2 - 2BC\cos(a)$$

$$\sin(x) = \cos(90° - x)$$

$$\sin(2x) = 2\sin(x)\cos(x)$$

$$\cos(2x) = 1 - 2\sin^2(x)$$

$$\sin(x + y) = \sin(x)\cos(y) + \cos(x)\sin(y)$$

$$\cos(x + y) = \cos(x)\cos(y) - \sin(x)\sin(y)$$

In order to get a better "feel" for the application of basic trigonometry, consider a disk rotating on a shaft. At the edge of the disk is a pin sticking out (like a handle). We will then place a "pickle fork" contraption over the pin in such a fashion that the fork can move up to down, but must remain horizontal in orientation:

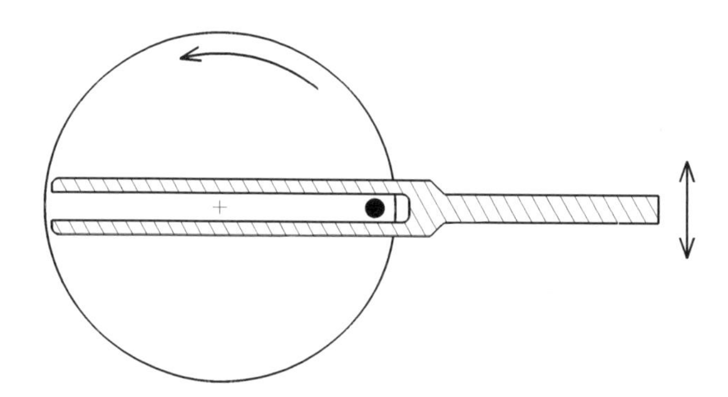

As the disk rotates on the shaft, the pickle fork will move up and down. When the pin is in the position shown, we will declare that the angle of the disk is zero, and we will mark the vertical position of the pickle fork as also being zero. As the disk rotates CCW, the pickle fork will move upward, then slow, and start downward. It will slow at the bottom, then return up to the starting position as the disk completes 360 degrees of rotation. The vertical motion of the pickle fork describes the sine of the angle of rotation of the pin.

If the pickle fork had been mounted at 90 degrees from where it is shown, its motion will describe the cosine of the angle of rotation.

The rotation of the disk is analogous to an electrical sinewave that is changing its phase (360 degrees per cycle as a function of time).

In looking more generally at a point on a circle, we can see relationships between the angle and the vertical component of its position (the sine), and the horizontal component of its position (the cosine):

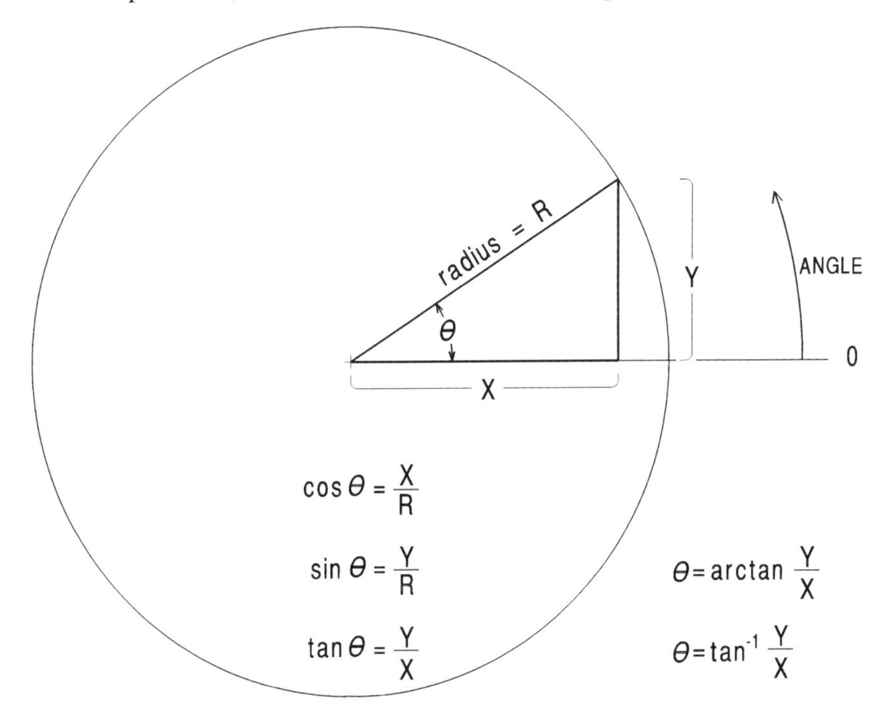

Knowing the vertical and horizontal components of the position of a point on the circle, you can determine the angle by using what is referred to as the "**arctangent**" of the ratio of the vertical to the horizontal components. This is simply the angle whose tangent is equal to the ratio. Arctangents (abbreviated as "arctan" or "atan", or "$\tan^{-1}$") can be determined from a table, or by using a calculator.

APPENDIX F- BIBLIOGRAPHY

ARRL publications:

- The Radio Amateur's Handbook *
- Advanced License Manual *
- Extra Class License Manual *
- Antenna Book
- Antenna Impedance Matching
- Electronics Data Book
- Microwave Handbook (volumes 1, 2, and 3)
- Microwave Experimenter's Manual
- UHF/Microwave Projects Manual
- Spread Spectrum Sourcebook
- Proceedings of Eastern States VHF/UHF Conferences
- Proceedings of Central States VHF Society Conferences
- Proceedings of West Coast VHF/UHF Conferences
- Proceedings of Microwave Update Conferences
- QEX Magazine (various issues)

RSGB Publications:

- Radio Data Reference Book
- UHF Compendium
- RadCom (various issues)

Spread Spectrum Systems, by Robert C. Dixon
Mobile Antenna Systems Handbook, by K. Fujimoto and J. R. James
Reflections, Transmission Lines, and Antennas, by Walter Maxwell, W2DU*
Foundations of Mobile Radio Engineering, by Michel Daoud Yacoub
Single Sideband Systems & Circuits (2nd edition) by William Sabin & Edgar Schoenike *
RF Design Magazine (various issues)
Filter Handbook Volumes 1 and 2, by RF Design Magazine
RF Circuit Design, by Chris Bowick *
The Art Of Electronics, by Paul Horowitz and Winfield Hill *
Electronic Design and Circuits, by Savant, Roden, and Carpenter
LC Filters, by Erich Christain
The Circuits and Filters Handbook, by Wai-Kai Chen
Reference Data For Radio Engineers, by ITT
The Electrical Engineering Handbook, by Richard C. Dorf
Electromagnetic Fields And Waves, by Magdy F. Iskander
Communications Receivers, by Bucher and Rohde
Antennas, by John D. Kraus
Transmission Line Transformers, by Jerry Sevick (W2FMI)
Electronic Designer's Handbook, by Landee, Davis, and Albrecht
Phaselock Techniques, by Floyd M. Gardner
Frequency Synthesis: Techniques And Applications, by Jerzy Gorski-Popiel
Frequency Synthesis Handbook. A collection of articles by RF Design Magazine
Information Transmission, Modulation & Noise, by Mischa Schwartz
An Introduction To Analog And Digital Communications, by Simon Haykin
Digital Communications, by Simon Haykin
Digital Communications, by Bernard Sklar
Principles of Secure Communication Systems, by Don J. Torrieri

Microwave Filters, Impedance-Matching Networks, & Coupling Structures, by Matthaei, Young, Jones.
Electronic Applications Of The Smith Chart, by Phillip H. Smith
High Frequency Amplifiers, by Ralph S. Carson
Filtering In The Time And Frequency Domains, by Blinchikoff and Zverev
Electronic Filter Design Handbook, by Arthur B. Williams
Active Network Design With Signal Filtering Applications, by Claude Lindquist
Active Inductorless Filters, by Sanjit K. Mitra
Stripline Circuit Design, by Harlan Howe
Foundations For Microstrip Circuit Design, by Terry Edwards
Microstrip Circuit Analysis, by David H. Schrader
Microstrip Circuits, by Fred Gardiol
Elements Of Information Theory, by Thomas M. Cover and Joy A. Thomas
Digital Filters, by Andreas Antniou
Microwave Engineering, by David M. Pozar
Plane Trigonometry and Statics, by Miller and Rourke

* - the references marked with an asterisk are recommended reading for someone wishing to rapidly gain an understanding of the subject matter without getting unduly involved in complex mathematics.

APPENDIX - G Sources of RF Equipment and Components

The following information lists just some of the available sources of RF Test Equipment, modules, and components. This is not by any means intended to be an endorsement of a particular manufacturer, or a recommendation of specific components or equipment.

Test Equipment:

Spectrum Analyzers

Keysight Technologies (formerly Agilent, formerly Hewlett Packard)
Anritsu
Tektronix
Rohde and Schwartz
Rigol
GW Instek

Vector Network Analyzers

Keysight Technologies
Anritsu
Rohde and Schwartz
Mini Radio Solutions

Synthesizers

Keysight Technologies
Anritsu
Rohde and Schwartz
Rigol
Mini-Circuits
GW Instek

Components:

Modular RF Amplifiers

Mini-Circuits
RF Parts
NXP Semiconductor (purchased the semiconductor division of Motorola)
MACOM

Mixers

MACOM
Mini-Circuits
IDT (Integrated Device Technology)
Digi-Key

Attenuators

Mini-Circuits
Pasternack Enterprises

RF Filters

Mini-Circuits
Pasternack Enterprises
Bird RF

RF Transistors

NXP Semiconductor
MACOM
ST Microelectronics
N-tronics
RF Parts
Digi-Key

27954532R00115

Made in the USA
Columbia, SC
02 October 2018